The Territories of Science and Religion

The Territories of

SCIENCE AND RELIGION

Peter Harrison

The University of Chicago Press
Chicago and London

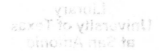

Peter Harrison is professor of the history of science and director of the Centre
for the History of European Discourses at the University of Queensland. He is
the author of numerous books, including *Wrestling with Nature: From Omens
to Science,* also published by the University of Chicago Press.

The University of Chicago Press, Chicago 60637
The University of Chicago Press, Ltd., London
© 2015 by The University of Chicago
All rights reserved. Published 2015.
Printed in the United States of America

24 23 22 21 20 19 18 17 16 15   1 2 3 4 5

ISBN-13: 978-0-226-18448-7   (cloth)
ISBN-13: 978-0-226-18451-7   (e-book)
DOI: 10.7208/chicago/9780226184517.001.0001

Library of Congress Cataloging-in-Publication Data
Harrison, Peter, 1955– author.
The territories of science and religion / Peter Harrison.
pages cm
Includes bibliographical references and index.
ISBN 978-0-226-18448-7 (cloth : alk. paper) —
ISBN 978-0-226-18451-7 (e-book)
1. Religion and science. 2. Science—History. I. Title.
BL240.3.H365 2015
201'.65—dc23

                           2014016857

∞ This paper meets the requirements of ANSI/NISO Z39.48-1992
(Permanence of Paper).

For Tom

[CONTENTS]

[PREFACE]

This book is a revised version of the Gifford Lectures, delivered at the University of Edinburgh in February 2011. The lectures sought to address the history of two overlapping sets of concerns—those to do with the nature of the physical universe and its operations, and those that concern the goals of human existence and the source of our moral values. We now tend to think of these questions as belonging to the distinct domains of science and religion. When we look to the past, however, we see that the boundaries of these two domains have been understood very differently and that questions concerning ultimate human meaning and value were rarely divorced from understandings of the nature of the universe. In one sense, then, this book is about the history of science and religion in the West. But it would be more accurate to say that it seeks to describe how it is that we have come to understand the world in terms of these distinct categories "science" and "religion"—how, in other words, we have come to separate the domain of material facts from the realm of moral and religious values.

This subject matter brings with it two particular challenges. The first should be more or less immediately apparent. This book cannot be a straightforward history of the relations between science and religion, since my argument is that these two ideas, as they are presently understood, are relatively recent conceptions that emerged in the West over the course of the past three hundred years. What I have sought to do, then, is to examine past activities that we have typically thought of in those terms, or have regarded as leading to them. A significant part of this exercise will be a consideration of the fortunes of the Latin terms *scientia* and *religio*. These two notions both begin as inner qualities of the individual—"virtues," if you will—before becoming concrete and abstract entities that are understood primarily in terms of doctrines and practices. This process of objectification is the precondition for a relationship between science and religion. In addition to a consideration of the Latin terms from which our modern English words "science" and "religion" derive, we shall also trace changing constellations of other conceptions that are genealogically related to our modern ideas of science and religion. They include "philosophy," "natural philosophy," "theology," "belief," and "doctrine," all of which had meanings for past historical actors that are quite unfamiliar to us today. One of my suggestions will be that there is a danger of systematically misconstruing past activities if we mistakenly assume the stability of meaning of these expressions.

A second challenge concerns the historical range of this book, which begins with classical Greece and early Christianity, and extends into the late nineteenth century. This may seem rather ambitious, particularly in light of a trend in the academy toward tight historical specialization. But this scope is necessary, in part because I am seeking to call into question common narratives about the trajectory of science that mirror this scale, beginning with its birth among the ancient Greeks, its decline in the Christian Middle Ages, its revival with the scientific revolution, and final triumph with the professionalization of science in the nineteenth century. It is also relevant that as a historical religion, modern Christianity still measures itself against its earliest forms, and this warrants paying attention to the early Christian era. That said, I have not attempted to provide a comprehensive history of the manifold ways in which, in the West, the study of the natural world was related to broader philosophical and religious concerns. Rather I have tried to "drill down" into particular historical moments, before moving on to conduct similar exercises for later historical periods. This comparative sampling approach will inevitably leave out some important parts of the story, but it does enable us to take stock of relevant notions at various points in history and make an assessment of the changes they have undergone. These gaps in the narrative also make it difficult to specify causes for all of the transitions that are identified, and in any case questions of historical causation are notoriously difficult. Nevertheless I will offer some suggestions about why what I consider to be the key transitions happen when they do.

The book is the culmination of a number of projects that I have been working on over the past twenty years, beginning with my early researches into our Western notion "religion" and taking in more recent work on the identity of philosophy and science in different historical periods and on the historical relations between science and religion. I will repeat versions of some arguments that I have advanced before, but the book offers what I believe is an entirely new perspective on these issues, particularly in its attempt to more closely relate the history of moral philosophy to the history of science. In relation to the lectures as delivered, while I have made significant revisions and additions, the six chapters of the book match the content of the six original lectures (in some cases more closely than in others). The first chapter offers a compressed overview of the general argument, and subsequent chapters fill out the details. I have also added scholarly apparatus in the form of fairly substantial endnotes for those interested in following up sources or who wish to pursue individual points in a little more detail. The notes will enable me to keep more arcane discussions out of the text and maintain at least some of the style of the oral presentations. In keeping with the original tone of the lectures on which this book is based, I have tried simply to tell the story, and have refrained (or at least have attempted to refrain) from intruding more recondite theoretical reflections into the narrative. That said, I have added a short epilogue in which I offer some brief remarks on theoretical matters along with observations on how my account relates to other discussions of the history of Western modernity.

[A NOTE ON THE GRAPHS]

Figure 3 is based on word searches in the Early English Books Online Text Creation Partnership.[1] Graphs of word searches covering the later modern period (figures 6, 9, 11–14) are based on Google Ngram viewer searches.[2] This technology, released in 2011, enables word-frequency searches in those texts currently in the Google Books database. The scale on the y axes is the percentage of titles in which the term occurs. The data are not entirely reliable, since the publication dates of some texts in the database have been incorrectly recorded. To the extent that it was possible, I have checked publication dates, particularly for earlier works, and have used the data to create bar graphs. I should add that I have used the graphs mostly for illustrative purposes, to confirm conclusions already established by other means.

[1]

The Territories of Science and Religion

How ridiculous are the boundaries of mortals!
—Seneca, *Natural Questions*[1]

The ideas entertained of God by wicked men must be bad, and
those of good men most excellent.
—Clement of Alexandria, *Stromata*[2]

Maps and Territories

If a historian were to contend that he or she had discovered evidence of
a hitherto unknown war that had broken out in the year 1600 between
Israel and Egypt, this claim would be treated with some skepticism. The
refutation of this claim would involve simply pointing out that the states
of Israel and Egypt did not exist in the early modern period, and that whatever conflicts might have been raging at this time could not on any reasonable interpretation be accurately described as involving a war between
Israel and Egypt. Neither would skeptical historians be impressed if their
colleague produced medieval maps that indicated the existence of such
places as Jerusalem and Alexandria, and included the various topographi-
cal features—rivers, deserts, mountains, plains, coastlines—that we currently might include in any description of the modern states of Israel

1

Figure 1. Abraham Ortelius's map of the Ottoman Empire, from the *Theatrum Orbis Terrarum* (1570).

and Egypt (see figure 1). At issue here would be not whether the relevant geographical territory existed then, but whether there were comparable boundaries and self-conscious national identities. Denial of the existence of a sixteenth-century Israel does not entail a denial of the existence of the territory that currently comprises that nation, but rather a denial that the territory was then viewed in a particular light, as something circumscribed by a set of boundaries and informed by particular ideals of nationhood.[3] During this period the territories of what we now know as Israel and Egypt were part of the same thing, namely, the Ottoman Empire. The idea of a medieval Israel and a medieval Egypt could only come about through the mistaken application of our present maps onto past territories.

My suggestion is that something similar is true for the entities "sci-

ence" and "religion," and more specifically, that many claims about putative historical relationships are confused for much the same reason as
claims about a sixteenth-century conflict between Israel and Egypt: that
is to say, they involve the distorting projection of our present conceptual
maps back onto the intellectual territories of the past. So familiar are the
concepts "science" and "religion," and so central to Western culture have
been the activities and achievements that are usually labeled "religious"
and "scientific," that it is natural to assume that they have been enduring
features of the cultural landscape of the West. But this view is mistaken.
To be sure, it is true that in the West from the sixth century BC attempts
were made to describe the world systematically, to understand the fundamental principles behind natural phenomena, and to provide naturalistic
accounts of the causes operating in the cosmos. Yet, as we shall see, these
past practices bear only a remote resemblance to modern science. It is also
true that almost from the beginning of recorded history many societies
have engaged in acts of worship, set aside sacred spaces and times, and
entertained beliefs about transcendental realities and proper conduct.
But it is only in recent times that these beliefs and activities have been
bounded by a common notion "religion," and have been set apart from the
"nonreligious" or secular domains of human existence.

In pointing out that "science" and "religion" are concepts of relatively
recent coinage, I intend to do more than make a historical claim about the
anachronistic application of modern concepts to past eras. What I have
in mind is not only to set out the story of how these categories "science"
and "religion" emerge in Western consciousness, but also to show how the
manner of their emergence can provide crucial insights into their present
relations. In much the same way that we can make sense of certain contemporary international conflicts by attending to the historical processes
through which national boundaries were carved out of a geographical territory, so too, with the respective territories of religion and the natural sciences. Just as the borders of nation-states are often more a consequence
of imperial ambitions, political expedience, and historical contingencies
than of a conscious attending to more "natural" fault lines of geography,
culture, and ethnicity—think in this context of the borders of the modern
state of Israel—so the compartmentalization of modern Western culture

that gave rise to these distinct notions "science" and "religion" resulted not from a rational or dispassionate consideration of how to divide cultural life along natural fracture lines, but to a significant degree has been to do with political power—broadly conceived—and the accidents of history.

The Joints of Nature

Another way of thinking about these two concepts is to consider an analogy with what philosophers call "natural kinds." The label "natural kind" is applied to natural groupings of things, the identity of which is natural in the sense that it does not depend on human beings. The sciences of chemistry and zoology, for example, seek to identify such kinds. Chemical elements and compounds are good examples of natural kinds—water, carbon, and hydrocarbons, for example. Occasionally, our everyday concepts, our ideas of what things go together do not map very well onto true natural kinds. We all know what jade is, for example. This lustrous, green, semiprecious stone would seem like a good candidate for being a natural kind. But as it turns out, there are two chemically distinct substances that are called "jade"—jadeite and nephrite.[4] One is a silicate of sodium and aluminum; the other a silicate of lime and magnesia. Once microstructure is taken into consideration, it becomes clear that jade is not a natural kind, because it is actually two different kinds of natural thing. My argument with regard to the categories "religion" and "science" is that to some degree we are mistaken in thinking that they are analogous to natural kinds, because despite the apparent similarities among those things that we call religions and the things that we call sciences, in fact the concepts and the way we deploy them masks important empirical differences.

In the case of jade, the assumption that there is a single unitary entity can be dispelled by careful measurement of some less obvious properties. The two minerals have slightly different specific gravities, refractive indices, and hardnesses. Infrared spectrographic analysis will also reveal their different chemical makeup. In the case of religion, my suggestion is that in addition to careful examination of the empirical characteristics of the so-called religions—which already brings to light an enormous and possibly irreconcilable diversity—their history is also revealing. Another instance of apparent kinds reinforces the point. Superficially, whales look like fish, and bats like birds, and folk taxonomies tend to group them together. Careful examination of the internal structures will reveal a different pattern of affinities, but so would the evolutionary history of these creatures, assuming that the latter could be established. The family histories of these groups would make it apparent that whales and bats should both properly be classified with the mammals. Similar considerations apply to both "religion" and "science," and we can reconstruct the history of these ideas with much greater precision than we can establish the phylogeny of biological taxa. What the history of the categories will show is how disparate, or at least significantly distinct, activities have come to be classified together. In the case of science, "natural history" and "natural philosophy" came together under the rubric "science" for the first time in the nineteenth century. These activities had involved quite different approaches to the study of nature and arguably their modern descendents—biology and physics—still exhibit the vestiges of their genealogical past. Thus, just as our use of the single word "jade" disguises the different composition of the two kinds that now bear that label, so the use of "science" for both historical sciences such as geology and evolutionary biology and physical sciences such as chemistry and physics tends to mask fundamental differences. These differences will necessarily complicate any global claims about the entities "science" and "religion" and their imagined relationship.

What follows from these considerations is that we distort the past if we uncritically apply our modern categories to past activities that would have been conceptualized by those who engaged in them in a quite different way. We should not use our present maps to understand their territory. We should not assume natural kinds where there are none. This means that the idea of a perennial conflict between science and religion must be false, just as claims about an early modern conflict between Israel and Egypt must be false. And this will be equally true for any claimed relationship between science and religion before the modern period. In addition we can say that contemporary "science-religion" relations, however construed (that is to say, whether positively or negatively), are to a large degree determined by the historical conditions under which disciplinary boundaries originated and developed over time. To advert once again to the map-territory analogy, we can ask whether the conceptual maps that we currently rely on to navigate through our cultural terrain are, to use the ugly but apt phrase, "fit for purpose." Thus, the question of the origins of boundaries can move beyond description and understanding to a critical inquiry into the appropriateness of how current conceptual maps divide territory. Good concepts, to use Plato's vivid image, carve nature at the joints (rather than, as he went on to say, dismembering it like a clumsy butcher).[5] Part of the burden of this book, then, is to ask whether these particular ways of dividing aspects of contemporary Western culture—"science" and "religion"—are helpful ones. In addressing this question I hope to show that "science" and "religion" are not self-evident or natural ways of dividing up cultural territory, that history shows this to be the case (as indeed does present consideration of cultures other than our own), and that persisting with these categories in an uncritical fashion can not only generate unhelpful conflict between science and religion, but can also disguise what perhaps ought to be legitimate sources of tension between the ways of faith and the formal study of nature. In short, this project attempts to set out a historical cartography of the categories "religion" and "science"—arguably the two cultural categories most important for an understanding of the nature of modernity and its legacy—with a view to casting light on their present relationship.

All of this implies that there is something not quite right with how we presently think about the relationship between science and religion, whether we think of it in terms of conflict or congruence, or even if we think that they do not have much to do with each other. Not only is too much of our present discussion uninformed by relevant historical considerations—imagine a comparable analysis of tensions in the Middle East that made no reference to history—but it is also often oblivious to the problematic nature of the categories in question. Much contemporary discussion about science and religion assumes that there are discrete human activities, "science" and "religion," which have had some unitary and enduring essence that persists over time. That this is not the case, I hope to illustrate in a number of ways, one of which involves closely attending to the history of the relevant terms.

In the remaining sections of this chapter I will make some rather cursory and compressed remarks about the history of the terms "religion" and "science" (or at least of their Latin equivalents). More extended treatments will come in the chapters that follow, but for now I am seeking simply to establish a basic case for the importance of reconsidering our historical understanding of these two concepts.

The History of "Religion"

In the section of his monumental *Summa theologiae* that is devoted to a discussion of the virtues of justice and prudence, the thirteenth-century Dominican priest Thomas Aquinas (1225–74) investigates, in his characteristically methodical and insightful way, the nature of religion. Along with North African Church Father Augustine of Hippo (354–430), Aquinas is probably the most influential Christian writer outside of the biblical authors. From the outset it is clear that for Aquinas religion (*religio*) is a virtue—not, incidentally, one of the preeminent theological virtues, but nonetheless an important moral virtue related to justice.[6] He explains that in its primary sense *religio* refers to interior acts of devotion and prayer, and that this interior dimension is more important than any outward expressions of this virtue. Aquinas acknowledges that a range of outward behaviors are associated with *religio*—vows, tithes, offerings, and so on—but he regards these as secondary. As I think is immediately obvious, this notion of religion is rather different from the one with which we are now familiar. There is no sense in which *religio* refers to systems of propositional beliefs, and no sense of different religions (plural).[7] Between Thomas's time and our own, *religio* has been transformed from a human virtue into a generic something, typically constituted by sets of beliefs and practices. It has also become the most common way of characterizing attitudes, beliefs, and practices concerned with the sacred or supernatural.

Aquinas's understanding of *religio* was by no means peculiar to him. Before the seventeenth century, the word "religion" and its cognates were used relatively infrequently. Equivalents of the term are virtually nonexistent in the canonical documents of the Western religions—the Hebrew Bible, the New Testament, and the Qur'an. When the term was used in the premodern West, it did not refer to discrete sets of beliefs and practices, but rather to something more like "inner piety," as we have seen in the case of Aquinas, or "worship." As a virtue associated with justice, moreover, *religio* was understood on the Aristotelian model of the virtues as the ideal middle point between two extremes—in this case, irreligion and superstition.[8]

The vocabulary of "true religion" that we encounter in the writings of some of the Church Fathers offers an instructive example. "*The* true religion" is suggestive of a system of beliefs that is distinguished from other such systems that are false. But careful examination of the content of these expressions reveals that early discussions about true and false religion were typically concerned not with belief, but rather worship and whether or not worship is properly directed. Tertullian (ca. 160–ca. 220) was the first Christian thinker to produce substantial writings in Latin and was also probably the first to use the expression "true religion." But in describing Christianity as "true religion of the true god," he is referring to genuine worship directed toward a real (rather than fictitious) God.[9] Another erudite North African Christian writer, Lactantius (ca. 240–ca. 320), gives the first book of his *Divine Institutes* the title "*De Falsa Religione*." Again, however, his purpose is not to demonstrate the falsity of pagan *beliefs,* but to show that "the religious *ceremonies* of the [pagan] gods are false," which is just to say that the objects of pagan worship are false gods. His positive project, an account of true religion, was "to teach in what manner or by what sacrifice God must be worshipped." Such rightly directed worship was for Lactantius "the duty of man, and in that one object the sum of all things and the whole course of a happy life consists."[10]

Jerome's choice of *religio* for his translation of the relatively uncommon Greek *thrēskeia* in James 1:27 similarly associates the word with cult and worship. In the English of the King James version the verse is rendered: "Pure and undefiled religion [*thrēskeia*] before God the Father is this, To visit the fatherless and widows in their affliction, and to keep himself unspotted from the world."[11] The import of this passage is that the "religion" of the Christians is a form of worship that consists in charitable acts rather than rituals. Here the contrast is between religion that is "vain" (*vana*) and that which is "pure and undefiled" (*religio munda et* inmaculata).[12] In the Middle Ages this came to be regarded as equivalent to a distinction between true and false religion. The twelfth-century *Distinctiones Abel* of Peter the Chanter (d. 1197), one of the most prominent of the twelfth-century theologians at the University of Paris, makes direct reference to the passage from James, distinguishing religion that is pure and true (*munda et vera*) from that which is vain and false (*vana et falsa*).[13] His pupil, the scholastic Radulfus Ardens, also spoke of "true religion" in this context, concluding that it consists in "the fear and love of God, and the keeping of his commandments."[14] Here again there is no sense of true and false doctrinal content.

Perhaps the most conspicuous use of the expression "true religion" among the Church Fathers came in the title of *De vera religion* (On True Religion), written by the great doctor of the Latin Church, Augustine of Hippo. In this early work Augustine follows Tertullian and Lactantius in describing true religion as rightly directed worship. As he was to relate in the *Retractions*: "I argued at great length and in many ways that true religion means the worship of the one true God."[15] It will come as no surprise that Augustine here suggests that "true religion is found only in the Catholic Church."[16] But intriguingly when writing the *Retractions* he was to state that while Christian religion is *a* form of true religion, it is not to be identified as *the* true religion. This, he reasoned, was because true religion had existed since the beginning of history and hence before the inception of Christianity.[17] Augustine addressed the issue of true and false religion again in a short work, *Six Questions in Answer to the Pagans*, written between 406 and 412 and appended to a letter sent to Deogratius, a priest at Carthage. Here he rehearses the familiar stance that true and false religion relates to the object of worship: "What the true religion reprehends in the superstitious practices of the pagans is that sacrifice is offered to false gods and wicked demons."[18] But again he goes on to explain that diverse cultic forms might all be legitimate expressions of true religion, and that the outward forms of true religion might vary in different times and places: "it makes no difference that people worship with different ceremonies in accord with the different requirements of times and places, if what is worshipped is holy." A variety of different cultural forms of worship might thus be motivated by a common underlying "religion": "different rites are celebrated in different peoples bound together by one and the same religion."[19] If true religion could exist outside the established forms of Catholic worship, conversely, some of those who exhibited the outward forms of Catholic religion might lack "the invisible and spiritual virtue of religion."[20]

This general understanding of religion as an inner disposition persisted into the Renaissance. The humanist philosopher and Platonist Marsilio Ficino (1433–99) thus writes of "christian religion," which is evidenced in lives oriented toward truth and goodness. "All religion," he wrote, in tones reminiscent of Augustine, "has something good in it; as long as it is directed towards God, the creator of all things, it is true Christian religion."[21] What Ficino seems to have in mind here is the idea that Christian religion is a *Christlike* piety, with "Christian" referring to the person of Christ, rather than to a system of religion—"*the* Christian religion." Augustine's suggestion that true and false religion might be displayed by Christians was also reprised by the Protestant Reformer Ulrich Zwingli, who wrote in 1525 of "true and false religion as displayed by Christians."[22]

It is worth mentioning at this point that, unlike English, Latin has no articles—no "a" or "the." Accordingly, when rendering expressions such as "*vera religio*" or "*christiana religio*" into English, translators had to decide on the basis of context whether to add an article or not. As we have seen, such decisions can make a crucial difference, for the connotations of "true religion" and "christian religion" are rather different from those of "*the* true religion" and "*the* Christian religion." The former can mean something like "genuine piety" and "Christlike piety" and are thus consistent with the idea of religion as an interior quality. Addition of the definite article, however, is suggestive of a system of belief. The translation history of Protestant Reformer John Calvin's classic *Institutio Christianae Religionis* (1536) gives a good indication both of the importance of the definite article and of changing understandings of religion in the seventeenth century. Calvin's work was intended as a manual for the inculcation of Christian piety, although this fact is disguised by the modern practice of rendering the title in English as *The Institutes of the Christian Religion*. The title page of the first English edition by Thomas Norton bears the more faithful "The Institution of Christian Religion" (1561).[23] The definite article is placed be-

fore "Christian" in the 1762 Glasgow edition: "The Institution of *the* Christian Religion." And the now familiar "Institutes" appears for the first time
in John Allen's 1813 edition: "The Institutes of the Christian Religion." The
modern rendering is suggestive of an entity "the Christian religion" that
is constituted by its propositional contents—"the institutes." These connotations were completely absent from the original title. Calvin himself
confirms this by declaring in the preface his intention "to furnish a kind
of rudiments, by which those who feel some interest in religion might be
trained to true godliness."[24]

With the increasing frequency of the expressions "religion" and "the
religions" from the sixteenth century onward we witness the beginning of
the objectification of what was once an interior disposition. Whereas for
Aquinas it was the "interior" acts of religion that held primacy, the balance now shifted decisively in favor of the exterior. This was a significant
new development, the making of religion into a systematic and generic
entity.[25] The appearance of this new conception of religion was a precondition for a relationship between science and religion. While the causes
of this objectification are various, the Protestant Reformation and the rise
of experimental natural philosophy were key factors, as we shall see in
chapter 4.

The History of "Science"

It is instructive at this point to return to Thomas Aquinas, because when
we consider what he has to say on the notion of science (*scientia*) we find
an intriguing parallel to his remarks on *religio*. In an extended treatment of
the virtues in the *Summa theologiae*, Aquinas observes that science (*scientia*) is a habit of mind or an "intellectual virtue."[26] The parallel with *religio*,
then, lies in the fact that we are now used to thinking of both religion and
science as systems of beliefs and practices, rather than conceiving of them
primarily as personal qualities. And for us today the question of their relationship is largely determined by their respective doctrinal content and
the methods through which that content is arrived at. For Aquinas, however, both *religio* and *scientia* were, in the first place, personal attributes.

We are also accustomed to think of virtues as belonging entirely within the sphere of morality. But again, for Aquinas, a virtue is understood more generally as a "habit" that perfects the powers that individuals possess.[27] This conviction—that human beings have natural powers that move them toward particular ends—was related to a general approach associated with the Greek philosopher Aristotle (384–322 BC), who had taught that all natural things are moved by intrinsic tendencies toward certain goals (telē). For Aristotle, this teleological movement was directed to the perfection of the entity, or to the perfection of the species to which it belonged.[28] As it turns out, one of the natural tendencies of human beings was a movement toward knowledge. As Aristotle famously wrote in the opening lines of the *Metaphysics*, "all men by nature desire to know."[29] In this scheme of things, our intellectual powers are naturally directed toward the end of knowledge, and they are assisted in their movement toward knowledge by acquired intellectual virtues.

One of the great revolutions of Western thought took place in the twelfth and thirteenth centuries, when much Greek learning, including the work of Aristotle, was rediscovered. Aquinas played a pivotal role in this recovery of ancient wisdom, making Aristotle one of his chief conversation partners. He was by no means a slavish adherent of Aristotelian doctrines, but nonetheless accepted the Greek philosopher's premise that the intellectual virtues perfect our intellectual powers. Aquinas identified three such virtues—understanding (*intellectus*), science (*scientia*), and wisdom (*sapientia*).[30] Briefly, understanding was to do with grasping first principles, science with the derivation of truths from those first principles, and wisdom with the grasp of the highest causes, including the first cause, God.[31] To make progress in science, then, was not to add to a body of systematic knowledge about the world, but was to become more adept at drawing "scientific" conclusions from general premises. "Science" thus understood was a mental habit that was gradually acquired through the rehearsal of logical demonstrations. In Thomas's words: "science can increase in itself by addition; thus when anyone learns several conclusions of geometry, the same specific habit of science increases in that man."[32]

These connotations of *scientia* were well known in the Renaissance and persisted until at least the end of the seventeenth century. The English physician John Securis wrote in 1566 that "science is a habit" and "a disposition to do any thing confirmed and had by long study, exercise, and use."[33] *Scientia* is subsequently defined in Thomas Holyoake's *Dictionary* (1676) as, properly speaking, the act of the knower, and, secondarily, the thing known. This entry also stresses the classical and scholastic idea of science as "a habit of knowledge got by demonstration."[34] French philosopher René Descartes (1596–1650) retained some of these generic, cognitive connotations when he defined *scientia* as "the skill to solve every problem."[35]

Yet, according to Aquinas, *scientia*, like the other intellectual virtues, was not solely concerned with rational and speculative considerations. In a significant departure from Aristotle, who had set out the basic rationale for an ethics based on virtue, Aquinas sought to integrate the intellectual virtues into a framework that included the supernatural virtues (faith, hope, and charity), "the seven gifts of the spirit," and the nine "fruits of the spirit."[36] While the various relations are complicated, particularly when beatitudes and vices are added to the equation, the upshot of it all is a considerable overlap of the intellectual and moral spheres. As philosopher Eleonore Stump has written, for Aquinas "all true excellence of intellect—wisdom, understanding and *scientia*—is possible only in connection with moral excellence as well." By the same token, on Aquinas's understanding, moral transgressions will have negative consequences for the capacity of the intellect to render correct judgments: "Carnal vices result in a certain culpable ignorance and mental dullness; and these in turn get in the way of understanding and *scientia*."[37] *Scientia*, then, was not only a personal quality, but also one that had a significant moral component.

The parallels between the virtues of *religio* and *scientia*, it must be conceded, are by no means exact. While in the Middle Ages there were no plural religions (or at least no plural religions understood as discrete sets of doctrines), there were undeniably sciences (*scientiae*), thought of as distinct and systematic bodies of knowledge.[38] The intellectual virtue *scientia* thus bore a particular relation to formal knowledge. On a strict definition, and following a standard reading of Aristotle's *Posterior Analytics*, a body of knowledge was regarded as scientific in the event that it had been arrived at through a process of logical demonstration.[39] But in practice the label "science" was extended to many forms of knowledge. The canonical divisions of knowledge in the Middle Ages—what we now know as the seven "liberal *arts*" (grammar, logic, rhetoric, arithmetic, astronomy, music, geometry)—were then known as the liberal *sciences*.[40] The other common way of dividing intellectual territory derived from Aristotle's classification of theoretical or speculative philosophy. In his discussion of the division and methods of the sciences, Aquinas noted that the standard classification of the seven liberal sciences did not include the Aristotelian disciplines of natural philosophy, mathematics, and theology. Accordingly, he argued that the label "science" should be given to these activities, too.[41] Robert Kilwardby (ca. 1215–79), successively regent at the University of Oxford and archbishop of Canterbury, extended the label even further in his work on the origin of the sciences, identifying forty distinct *scientiae*.[42]

The English word "science" had similar connotations. As was the case with the Latin *scientia*, the English term commonly referred to the subjects making up the seven liberal arts. In catalogs of English books published between 1475 and 1700 we encounter the natural and moral sciences, the sciences of physick (medicine), of surgery, of logic and mathematics. Broader applications of the term include accounting, architecture, geography, sailing, surveying, defense, music, and pleading in court. Less familiarly, we also encounter works on the science of angels, the science of flattery, and in one notable instance, the science of drinking, drolly designated by the author the "eighth liberal science."[43] At nineteenth-century Oxford "science" still referred to elements of the philosophy curriculum.[44] The idiosyncrasies of English usage at the University of Oxford notwithstanding, the now familiar meaning of the English expression dates from the nineteenth century, when "science" began to refer almost exclusively to the natural and physical sciences.

Returning to the comparison with medieval *religio*, what we can say is that in the Middle Ages both notions have a significant interior dimension, and that what happens in the early modern period is that the balance between the interior and exterior begins to tip in favor of the latter. Over the course of the sixteenth and seventeenth centuries we will witness the beginning of a process in which the idea of religion and science as virtues or habits of mind begins to be overshadowed by the modern, systematic entities "science" and "religion." In the case of *scientia,* then, the interior qualities that characterized the intellectual virtue of *scientia* are transferred to methods and doctrines. The entry for "science" in the 1771 *Encyclopaedia Britannica* thus reads, in its entirety: "SCIENCE, in philosophy, denotes any doctrine, deduced from self-evident and certain principles, by a regular demonstration."[45] The logical rigor that had once been primarily a personal characteristic now resides primarily in the corresponding body of knowledge.

The other significant difference between the virtues of *religio* and *scientia* lies in the relation of the interior and exterior elements. In the case of *religio,* the acts of worship are secondary in the sense that they are motivated by an inner piety. In the case of *scientia,* it is the rehearsal of the processes of demonstration that strengthens the relevant mental habit. Crucially, because the primary goal is the augmentation of mental habits, gained through familiarity with systematic bodies of knowledge ("the sciences"), the emphasis was less on the production of scientific knowledge than on the rehearsal of the scientific knowledge that already existed. Again, as noted earlier, this was because the "growth" of science was understood as taking place within the mind of the individual. In the present, of course, whatever vestiges of the scientific *habitus* remain in the mind of the modern scientist are directed toward the production of new scientific knowledge. In so far as they exist at all—and for the most part they have been projected outward onto experimental protocols—they are a means and not the end. Overstating the matter somewhat, in the Middle Ages scientific knowledge was an instrument for the inculcation of scientific habits of mind; now scientific habits of mind are cultivated primarily as an instrument for the production of scientific knowledge.

The atrophy of the virtues of *scientia* and *religio,* and the increasing emphasis on their exterior manifestations in the sixteenth and seventeenth centuries, will be discussed in more detail in chapter 4. But looking ahead we can say that in the physical realm virtues and powers were removed from natural objects and replaced by a notion of external law. The order of things will now be understood in terms of laws of nature—a conception that makes its first appearance in the seventeenth century—and these laws will take the place of those inherent tendencies within things

that strive for their perfection. In the moral sphere, a similar development takes place, and human virtues will be subordinated to an idea of
divinely imposed laws—in this instance, moral laws. The virtues—moral
and intellectual—will be understood in terms of their capacity to produce
the relevant behaviors or bodies of knowledge. What drives both of these
shifts is the rejection of an Aristotelian and scholastic teleology, and the
subsequent demise of the classical understanding of virtue will underpin
the early modern transformation of the ideas of *scientia* and *religio*.

Science and Religion?

It should by now be clear that the question of the relationship between
science (*scientia*) and religion (*religio*) in the Middle Ages was very different from the modern question of the relationship between science and
religion. Were the question put to Thomas Aquinas, he may have said
something like this: Science is an intellectual habit; religion, like the
other virtues, is a moral habit.[46] There would then have been no question
of conflict or agreement between science and religion because they were
not the kinds of things that admitted those sorts of relations. When the
question is posed in our own era, very different answers are forthcoming, for the issue of science and religion is now generally assumed to be
about specific knowledge claims or, less often, about the respective processes by which knowledge is generated in these two enterprises. Between
Thomas's time and our own, *religio* has been transformed from a human
virtue into a generic something typically constituted by sets of beliefs and
practices. *Scientia* has followed a similar course, for although it had always
referred both to a form of knowledge and a habit of mind, the interior dimension has now almost entirely disappeared. During the sixteenth and
seventeenth centuries, both religion and science were literally turned in-
side out.

Admittedly, there would have been another way of posing this question
in the Middle Ages. In focusing on *religio* and *scientia* I have considered the
two concepts that are the closest *linguistically* to our modern "religion" and
"science." But there may be other ancient and medieval precedents of our

modern notions "religion" and "science," that have less obvious linguistic connections.[47] It might be argued, for example, that two other systematic activities lie more squarely in the genealogical ancestry of our two objects of interest, and they are *theology* and *natural philosophy.* A better way to frame the central question, it could then be suggested, would be to inquire about theology (which looks very much like a body of religious knowledge expressed propositionally) and natural philosophy (which was the name given to the systematic study of nature up until the modern period), and their relationship.[48]

There is no doubt that these two notions are directly relevant to our discussion, but I have avoided mention of them up until now, first, because I have not wished to pull apart too many concepts at once and, second, because we will be encountering these two ideas and the question of how they fit into the trajectory of our modern notions of science and religion in subsequent chapters. For now, however, it is worth briefly noting that the term "theology" was not much used by Christian thinkers before the thirteenth century. The word *theologia* appears for the first time in Plato (ca. 428–348 BC), and it is Aristotle who uses it in a formal sense to refer to the most elevated of the speculative sciences.[49] Partly because of this, for the Church Fathers "theology" was often understood as referring to *pagan* discourse about the gods. Christian writers were more concerned with the interpretation of scripture than with "theology," and the expression "sacred doctrine" (*sacra doctrina*) reflects their understanding of the content of scripture. When the term does come into use in the later Middle Ages, there were two different senses of "theology"—one a speculative science as described by Aristotle, the other the teaching of the Christian scriptures.[50]

Famously, the scholastic philosophers inquired as to whether theology (in the sense of *sacra doctrina*) was a science. This is not the place for an extended discussion of that commonplace, but the question does suggest one possible relation between science and theology—that theology is a species of the genus "science." Needless to say, this is almost completely disanalogous to any modern relationship between science and religion as we now understand them. Even so, this question affords us the opportu-

nity to revisit the relationship between virtues and the bodies of knowledge that they were associated with. In so far as theology was regarded
as a science, it was understood in light of the virtue of *scientia* outlined
above. In other words, theology was also understood to be, in part, a mental habit. When Aquinas asks whether sacred doctrine is one science,
his affirmative answer refers to the fact that there is a single faculty or
habit involved.[51] His contemporary, the Franciscan theologian Bonaventure (1221–74), was to say that theological science was a habit that had
as its chief end "that we become good."[52] The "subtle doctor," John Duns
Scotus (ca. 1265–1308), later wrote that the "science" of theology perfects
the intellect and promotes the love of God: "The intellect perfected by the
habit of theology apprehends God as one who should be loved."[53] While
these three thinkers differed from each other significantly in how they
conceptualized the goals of theology, what they shared was a common
conviction that theology was, to use a current expression somewhat out
of context, habit forming.

As for "natural philosophy" (*physica, physiologia*), historians of science
have argued for some years now that this is the closest ancient and medieval analogue to modern science, although they have become increasingly sensitive to the differences between the two activities.[54] Typically,
these differences have been thought to lie in the subject matter of natural
philosophy, which traditionally included such topics as God and the soul,
but excluded mathematics and natural history. On both counts natural
philosophy looks different from modern science. What has been less well
understood, however, are the implications of the fact that natural philosophy was an integral part of philosophy. These implications are related to
the fact that philosophy, as practiced in the past, was less about affirming
certain doctrines or propositions than it was about pursuing a particular
kind of life. Thus natural philosophy was thought to serve general philosophical goals that were themselves oriented toward securing the good
life. These features of natural philosophy will be discussed in more detail
in the chapter that follows. For now, however, my suggestion is that moving our attention to the alternative categories of theology and natural philosophy will not yield a substantially different view of the kinds of historical transitions that I am seeking to elucidate.

::::

The nineteenth-century French scholar Ernest Renan (1823–92), while probably best known for his popular and controversial *Life of Jesus* (1863), also penned the classic essay "What Is a Nation?" (1882). Here, he offers this telling remark: "Forgetting—I would go so far as to say historical error—is a crucial factor in the creation of a nation."[55] The specific case that Renan had in mind was the reified object "*la France.*" France, he suggests, invented nationhood, and all other models of the nation were henceforth regarded as inferior imitations. But the very idea of France required the jettisoning of a burden of unwelcome historical truths. While we are all aware of the manner in which the idea of a nation involves founding myths of various kinds, and through them, the positive construction of a particular ideal, we are here reminded that founding myths also require a kind of negation—an amnesia about what came before, and a forgetting of historical realities that might challenge the integrity of our new conception. Indeed, Karl Deutsch's similarly unflattering definition of a nation— "a group of people united by a mistaken view about the past and a hatred of their neighbours"—is not an altogether unfitting description for those who in recent times have sought to foment hostility between science and religion.[56] This first chapter has provided some reasons for thinking that the creation of the modern boundaries of these concepts has also been accompanied by a kind of historical amnesia. In the chapters to come, I hope to expose some of the myths that inform our present categories, and to offer further details of an alternative, and largely forgotten history—a history that, once called to mind, may help us reconfigure the relationship between the entities that we now call "science" and "religion."

[2]

The Cosmos and the Religious Quest

Natural philosophy substitutes for festering superstition that
unshaken piety that is attended by good hopes.
—Plutarch, *Life of Pericles*[1]

It is not a question of a "doctrine" being handed down by uniform
repetition or arbitrarily distorted; it is a question of a life, again
and again kindled afresh, and now burning with a flame of its own.
—Adolf Harnack, *What Is Christianity?*[2]

If science is a modern idea, we must ask ourselves what those whom we
traditionally regard as having pursued science in antiquity imagined
themselves to be doing. Equally, if there was no religion before the early
modern period, how did such groups as the early Christians conceptualize what it was that they were committed to? These are the two questions
that will be explored in this chapter. As I have suggested, the historical reconstruction of these earlier activities calls for a recapturing of forgotten
ways of life, along with a dispelling of those myths upon which the modern boundaries of science and religion have been constructed. We begin
with one such myth about the origins of science in antiquity.

The Mythical Origins of Science

The history of science, on one very common understanding, has three distinct stages. Science is said to have had its origins in Greek antiquity when philosophers first broke away from the myths of their forebears and sought rational explanations for natural phenomena. Science subsequently suffered a setback with the advent of Christianity, going into significant decline in the Middle Ages. But it then emerged triumphant with the scientific revolution of the seventeenth century when it finally broke away from religion and set out on its progressive path to the present.

In this version of events the honor of having founded Western science is usually conferred upon the philosopher Thales (d. 546 BC), who lived in the port city of Miletus, a Greek colony on the west coast of what is now Turkey.[3] The characteristic features of his thought, on account of which he is so honored, include the rejection of supernatural explanations, a search for unitary natural principles, and a willingness to engage in rational debate about the relative merits of various speculations about the world and its operations. Western science was thus said to have been born among the ancient Greeks and to have been developed by them to a state of some sophistication. With the fall of Rome and the rise of Christianity, however, the fortunes of science waned. The leading intellects in medieval Christendom directed their mental energies toward theology and had little time for the systematic study of nature. Then, with the Renaissance and scientific revolution, the aims and ideals of ancient science were reborn. Although the scientific doctrines of such seventeenth-century figures as Galileo, Boyle, and Newton were different in content from anything that had come before, these men were regarded as having investigated nature with the same spirit that had motivated the endeavors of their Greek forebears. These luminaries laid new foundations for the sciences, and on these foundations rest the remarkable accomplishments of subsequent generations of scientists in a lineage that extends to the present.

The key moments of transition in these large-scale histories are the dawn of science among the ancient Greeks, its decline with the accession of Christianity, and its rebirth in the early modern period with the revival of the ideals of the ancients. The scientific revolution of the seventeenth century is the pivotal event in these histories, for it not only looks back to the ethos of the ancient Greek investigators of nature, but also anticipates the impressive achievements of the modern sciences. The eminent Cambridge historian Herbert Butterfield had this vision of history in mind when he remarked that the scientific revolution "outshines everything since the rise of Christianity and reduces the Renaissance and Reformation to the rank of mere episodes." As for the progenitors of science, Butterfield was happy to concede that "natural science itself came to the modern world as a legacy from ancient Greece."[4]

One of the interesting characteristics of this view of science and its history is that from its very inception, science is placed in a particular relation to religion. What makes the speculations of Thales and his Milesian school "scientific" is that they are imagined to represent a definitive break with prevailing mythological or religious explanations of natural phenomena.[5] In these narratives, science, even at its birth, is distinguished by its capacity to provide alternative and more rational accounts of the cosmos than those offered by religion and mythology. The second phase of these histories similarly understands science and religion as being related in a specific way. Classical culture is often imagined to have somehow "lost its nerve," degenerating into an "age of anxiety" that paved the way for the rise of mystery religions and the eventual success of Christianity.[6] For their part, patristic and medieval Christian authors were said to have associated Greek science with paganism and thus to have discouraged its practice. Their attentions were firmly fixed not on the physical operations of the present world, but on fulfilling the requirements necessary for salvation in the next. As a consequence, science is said to have been either hindered or ignored by the Church Fathers and their scholastic successors in the medieval age of faith. By implication, the scientific revolution was accomplished only by overcoming the religious preoccupations and prejudices of the previous age. The notorious example of Galileo's confrontation with the Inquisition seems to confirm this view of things, and the fact that the rise of science in the modern period was followed by the secularization of European society counts as further evidence of a generally negative relation between science and religion.

While historians of science have now largely abandoned much of this

narrative, it continues to exercise a tenacious hold on the popular imagination and still informs many nonspecialist accounts of science and its
history. In one of his last essays, the influential philosopher of science Karl
Popper (1902–94) asserted that "the scientific tradition" was inaugurated
by Thales and his immediate successors, and that it died in the West when
it was suppressed by "a victorious and intolerant Christianity." While science was "missed and mourned during the Middle Ages," it was eventually
revived during the Renaissance and "found fulfilment in Newton."[7] Similar contentions may also be found in many popular works on science and
in school textbooks. The physicist and science writer Paul Davies declares
in one of his recent books that religion was "the first systematic attempt to
explain the universe." But then came science, which cast its explanations
"in terms of impersonal forces and natural physical processes rather than
the activities of purposeful supernatural agents." Davies goes on to explain
that whenever religious and scientific explanations came into conflict, invariably it was science that emerged victorious.[8] In Brian Bunch's recent
survey of the history of science and technology we are again informed
that Western science began with Thales of Miletus. He and his successors
"were the first to believe that the universe could be understood using reason alone rather than through mythology or religion." As a consequence
of these ideas, the ancient Greeks established institutions in which individuals "pursued science in somewhat the way the universities do today."
The advent of Christianity in late antiquity brought an end to these institutions, but science was revived during the scientific revolution.[9]

Comforting though this narrative may be for some, the reality is rather
different. If we focus for now simply on the role ascribed to Greek science, we can say that its putative rejection of myth and supposed incompatibility with religion break down under close scrutiny. No one who has
read the extant fragments of the pre-Socratic philosophers can fail to be
struck by their ubiquitous references to gods and divine principles. Thales,
the purported progenitor of science, declared that "all things are full of
gods," and on discovering his famous theorem he is said to have sacri-
ficed an ox.[10] These are not the actions of a hard-nosed scientific naturalist. Anaxagoras (b. ca. 500 BC), like Thales, is often portrayed as exemplifying a scientific naturalism that is essentially incompatible with

a theological understanding of the cosmos. This characterization draws some credibility from the claim that Anaxagoras was banished from Athens on account of his skeptical claims that the sun was just a mass of molten metal, that the moon was made of an earthlike substance, and that the stars were merely fiery stones.[11] But it was this same Anaxagoras who contended that the whole universe was controlled by a divine causal principle (nous—mind or intellect), a view that was to influence, in various ways, Plato, Aristotle, the Stoics, and the Neoplatonists, and which came to underpin much of the subsequent ancient Greek belief in the inherent rationality of the natural world. Indeed, in various ways other pre-Socratic philosophers had postulated similar principles—Anaximander's *Apeiron*, Heraclitus's *Logos*, Xenophanes's "One God"—that imply an ordered, yet divinely animated, cosmos.[12]

The notion of distinct and successive mentalities—mythopoeic and rationalist—is also difficult to sustain. It is striking, for example, that the beginnings of (scientific) Hippocratic medicine coincide exactly with the rise of the (religious) cult of Asclepius, and that these apparently incompatible approaches to healing coexist happily in the same geographical regions. Practitioners of these distinct therapies also shared many of the same methods, to say nothing of the fact that the Hippocratic oath invokes Apollo, Asclepius, and "all the gods and goddesses."[13] To dismiss this as pious window dressing is to fail to understand how the religious perspective then pervaded every area of life. The narrative of an opposition between "science" and myth also betrays too crude an understanding of the role of myth and its relation to reason. Myths were not thought to offer alternative explanatory accounts to "science." Not only were they regarded as compatible with rational, philosophical accounts of the natural world, but they were also considered to be important vehicles for the transmission of profound philosophical truths.[14] It is a mistake, then, to regard myths as incompatible with rational explanations, or to imagine that a mythical phase of Western history gave way to a proto-scientific age.

But perhaps the most significant deficiency in this common reconstruction of the history of science lies in the assumption that these ancient Greek accounts of the cosmos partake of the ethos of modern science, and that they share to a significant degree its goals and methods.

Natural Philosophy and the Good Life

The ancient Greeks had neither activities nor occupations that are directly equivalent to our terms "science" and "scientist."[15] Those who concerned themselves with the phenomena of nature were then known as "natural philosophers" (*physici*) and their activities fell under the rubric of philosophy. While it is certainly possible to project back modern notions of science onto the activities of the ancients, to do so uncritically does violence to the relations of natural philosophy with other aspects of ancient culture such as poetry, mythology, and religion. Natural philosophy (or "physics") was an integral part of philosophy itself. However, this realization is only helpful if we understand that ancient philosophy has only the most tenuous connection with the subject matter now taught in university departments of philosophy (and particularly those that cleave to the analytic tradition). Philosophy then entailed the pursuit of wisdom or happiness: it was, to use the succinct expression of philosopher Pierre Hadot, "a way of life."[16] The doctrinal content of philosophy was subservient to the art of living.[17]

Socrates (ca. 469–399 BC) thus engages in arguments and pursues definitions, but with the aim of promoting the philosophical life. As related in vivid detail in the *Apology,* Socrates, on trial for his life, offers a compelling defense of his philosophical mission. He explains that philosophy is the business of examining life and that its aim is the moral reform of the individual.[18] Elsewhere he declares that the philosophical quest is directed to the discovery of "what kind of life one should live."[19] The ancient philosophical schools, for all their differences, agreed that philosophy was about how life was to be lived. The much misunderstood Epicurus (341–270 BC) could thus define philosophy as the activity "which by words and arguments secures the happy life."[20] For the Epicureans philosophy did involve words and arguments, but these were but a means of securing happiness. The Stoics took a similar view, and the Roman philosopher and statesman Seneca (ca. 3 BC–AD 65) found it necessary to remind his contemporaries that philosophy was not simply a matter of being familiar with doctrines or dialectical techniques. Instead, he maintained, philosophy "moulds and constructs the soul; it orders our life, guides our conduct, shows us what we should do and what we should leave undone."[21] His successor in the line of Stoic philosophers, Epictetus (AD 55–135), similarly chided those "who embrace philosophy only in words." He argued that philosophy does not promise anything "external for man," for it concerns "the art of living" and its raw material is the individual's own life.[22] The skeptics, of all of the philosophical schools, perhaps represent the ultimate subordination of doctrines to the living of the philosophical life. Confronted with the conflicting propositional claims of competing philosophical traditions, the skeptic recognizes the need to suspend judgment (*epochē*). This in turn leads to tranquility of mind (*ataraxia*).[23] Paradoxically, for the skeptics the quest for truth is necessarily followed by the abandonment of doctrinal philosophy, and this leads in turn to the desired philosophical state.[24]

One general question to arise out of this identification of ancient philosophy with the art of living is to do with the relationship between the doctrinal content of philosophy—which modern philosophers have tended to regard as constitutive of ancient philosophy—and the good life.[25] Related to this is the more specific, and more puzzling, question of how the study of nature—natural philosophy—might have contributed to philosophy proper, understood as the art of living. How, for example, could an understanding of the motions of the heavens promote the moral goals of the philosophical way of life? A third question, granting all this, is how over the course of two millennia the study of nature came to be dissociated from the moral realm, and how a modern science developed that is devoid of ethical content and largely indifferent to the moral commitments of its practitioners. Some suggestions about this third question will be forthcoming, but the first two are our immediate concern.

Socrates, it must be said, seems to have held that the study of the cosmos was largely irrelevant to what really mattered to the philosopher.[26] But for a number of his successors a connection between the study of nature and the philosophical life was provided by the assumption that a moral order is built into the structure of the cosmos.[27] The various divine principles identified by the pre-Socratic philosophers were to promote the widespread assumption that for those engaged in the philosophical quest, the universe provides a model of the good. To take a key example, in the

Republic Plato held that the philosopher must engage in contemplation of the heavenly order in which "all things move according to reason" so that he might conform to that same pattern and lead a life that was "orderly and divine."[28] In the *Timaeus*—a work that should remind us of how myth and philosophy can work together—this idea is developed further:

> And the motions which are naturally akin to the divine principle within us are the thoughts and revolutions of the universe. These each man should follow, and by learning the harmonies and revolutions of the universe, should correct the courses of the head which were corrupted at our birth, and should assimilate the thinking being to the thought, renewing his original nature, so that having assimilated them he may attain to that best life which the gods have set before mankind, both for the present and the future.[29]

For Plato, it was the *mathematical* study of the heavens in particular that contributed to the moral and intellectual formation of the philosopher. He thus contended that mathematics is a kind of "divine art" that raises the mind to a godlike state: hence the tradition according to which the entrance to Plato's Academy bore the inscription: "Let no-one ignorant of geometry enter here."[30]

For all his differences with Plato, Aristotle agreed that the cosmos was characterized by a particular harmony and order, and that the natural parts of human beings correspond to those of the universe.[31] Like Plato, he also associated mathematics with "the true and the good."[32] And he contended that the goal of philosophy was to secure the kinds of moral attributes that would make one "supremely happy."[33] The Stoics famously regarded the good life as a life lived in accordance with nature. Zeno of Elea (ca. 490–ca. 430 BC), the founder of Stoicism, declared that the virtuous life was the same as "life in agreement with nature."[34] "Live according to nature" was a predominant theme in the letters of Seneca.[35] Epictetus likewise explained that God had introduced us into the world as both spectators and interpreters of his works. The end of life lay "in contemplation and understanding, and in a scheme of life conformable to nature."[36]

Knowledge of nature thus enabled the philosopher to align his life with the rational principle that pervaded the cosmos.

If we inquire after the specific mechanisms through which doctrines about the natural world might relate to philosophical practices, we have good examples in both the Stoics and the Epicureans. For the Stoics, living in accordance with nature required a knowledge of nature and its operations. One reason for this was that the study of nature was thought to offer the best way of establishing what lay within one's own power, and what in the power of nature. This, in turn, enabled the individual to discern what could be changed, and what was to be accepted with equanimity. As Epictetus explains:

> All other creatures are indeed excluded from a power of comprehending the administration of the world; but a reasonable being . . . sees that of the things which relate to it some are unrestrained and in its own power, some restrained and in the power of others: the unrestrained, such as depend on will; the restrained, such as do not depend on it. And for this reason, if it esteems its good and its interest to consist in things unrestrained and in its own power, it will be free, prosperous, happy, safe, magnanimous, pious, thankful to God for everything, never finding fault with anything, never censuring anything that is brought about by him.[37]

In other words, while we cannot control the course of nature, it is within our power to control our own attitude to natural events. Knowledge of natural causes thus provides a buffer against the vicissitudes of life, making possible genuine happiness. The renowned nineteenth-century historian Thomas Babington Macaulay rightly observed that the Stoic philosopher Seneca had concerned himself with natural philosophy and had "magnified the importance of its study." Why, he asks? "Not because it tended to assuage suffering, to multiply the conveniences of life, to extend the empire of man over the material world; but solely because it tended to raise the mind above low cares, to separate it from the body, to exercise its subtlety in the solution of very obscure questions." Natural philosophy, he concludes, was a mere "mental exercise."[38]

30 Chapter Two

The Epicureans also offer a clear model of how the study of nature was conducive to the good life, and this in spite of the fact that unlike the Stoics, they did not believe that a moral order pervaded the cosmos. For Epicurus, the study of nature was important because once we understand the true nature of things we will fear neither death nor the prospect of postmortem punishment, and hence will be free to pursue genuine happiness.[39] The aim of natural philosophy is to free us from irrational fears. In Cicero's *De finibus*, the spokesman for Epicureanism thus declares that "knowledge of the facts of nature relieves us of the burden of superstition, frees us from fear of death, and shields us against the disturbing effects of ignorance which is often in itself a cause of terrifying apprehensions."[40] Epicurus himself had thought that one practical remedy for such fears was the rehearsal of a memorized "catechism," the content of which was the doctrines of physics: "Since such a course is of service to all who take up natural science [*physiologia*], I who devote to the subject my continuous energy and reap the calm enjoyment of a life like this have prepared for you just such an epitome and manual of the doctrines as a whole."[41] These sovereign doctrines, once committed to memory, were to become second nature. They would stabilize the delicate atoms of the soul and prevent them from being swayed by images and impressions from without.[42]

The Epicurean use of physics in this way offers us a good example of what Hadot refers to as "spiritual exercises" (*áskēsis*). While the term is familiar perhaps, from the title of Ignatius of Loyola's well-known *Exercitia spiritualia* (1548)—a manual of contemplation, meditation, and self-review, set in the context of Catholic spirituality—Hadot argues that the expression is the best way to characterize the therapeutic philosophical practices of the Greco-Roman philosophical schools: "Each school had its own therapeutic method, but all of them linked their therapeutics to a profound transformation of the individual's mode of seeing and being. The object of spiritual exercises is precisely to bring about this transformation."[43] When Christian writers subsequently speak of exercises devoted to the cultivation or cure of the soul, they are simply carrying over this classical idea. To take just one example, Ambrose of Milan (ca. AD 340–97), one of the great doctors of the Western church, stated that "forti-

tude of the soul is enhanced by a course of exercises." These exercises consisted in prayer, meditation, and the reading of scripture.[44]

The formative element of natural philosophy is also apparent in the
various classifications of knowledge observed by the ancients. In his influential taxonomy, Aristotle had distinguished three speculative or theoretical sciences: natural philosophy, mathematics, and theology. (Ethics
was separately classified as a "practical science"; the arts were "productive
sciences.")[45] Aristotle's speculative sciences were distinguished by their
respective objects: *theology*, the most elevated of the sciences, deals with
what is eternal, immovable, separable from matter; *natural philosophy* is
concerned with objects that are the opposite in each respect, that is, finite, movable, and inseparable from matter. The objects of *mathematics*
(or at least some parts of it) lie between these, being immovable, probably
not separable from matter, but embodied in it.[46] The tendency of modern
scholars, when considering these classifications, has often been to look
upon them as defining what we would call "disciplinary boundaries" —
distinguishing the subject matters of the various academic activities. But
these classifications were not merely ways of dividing up theoretical content: they were intended to set out the proper progression in the order
of study, beginning with the material and mutable (natural philosophy),
proceeding then to the immutable things that were associated with material objects (mathematics), before moving on to elevated divine things
that were immutable and completely divorced from matter. The habits of
mind developed in the more materially oriented sciences would naturally
prepare the mind for the kinds of mental transformations demanded by
the more elevated sciences.[47]

Xenocrates, Plato, the Stoics, and the Epicureans had adopted a different way of dividing philosophical subject matter from Aristotle, speaking
instead of physics (or natural philosophy), logic, and ethics.[48] But again
the aim of study was personal formation. The Stoics believed that the two
sciences of physics and logic ultimately served a third science, that of
ethics. Theology, incidentally, they held to be the most elevated branch of
physics. In Stoic philosophy those who understand the operations of the
cosmos will also know how to conduct themselves within the cosmos.[49]

Epicurus spelled out how the various branches of knowledge shape the mental habits of the philosopher. Thus logic was pursued in order to discipline judgment, physics to discipline desire, and ethics to discipline the inclinations.[50]

This tendency to fuse "scientific," ethical, and religious concerns became even more pronounced in the philosophical schools of late antiquity. The Neoplatonist philosopher Porphyry (AD 234–ca. 305) was typical in his insistence that philosophy was to do with neither "the accumulation of arguments" nor the profession of "learned knowledge," but about transforming one's own life.[51] Porphyry's disciple Simplicius (ca. 490–560) offered a similar view of progression from the study of nature to the higher activities of contemplation and religious piety:

Physics (phusiologia) is useful: in the affairs of daily life, because it provides the principles of technologies such as medicine and mechanics (understood above all as the art of manufacturing machines of war) because it contributes to leading the superior part of the soul, which is the intellect, towards its perfection—for which study of theology is particularly valuable; it is an auxiliary for moral virtues; a ladder that leads towards knowledge of God and ideas; and finally it incites us to piety and to acts of thanksgiving towards God.[52]

Simplicius, we should note, was no friend to Christianity, in spite of these pious sentiments. It is also significant that Simplicius should speak in terms of the "usefulness" of physics, and that he lists the uses of physics in order of increasing importance. An emphasis on the utility of the sciences has always been part of their justification, but as we shall see in chapter 5, what counts as utility changes over time. The modern period will witness the beginning of an inversion of the priorities of Simplicius, when technological applications and the affairs of daily life begin to take precedence over moral edification. Simplicius's image of the ladder also reminds us that physics was often understood as the first stage of the ascent of the mind to God.

Finally, offering an explicit link to the "scientific" cosmology most closely identified with antiquity and the Middle Ages, is Claudius Ptolemy (ca. AD 90–168). In his best-known work, the *Almagest*—which sets out with remarkable ingenuity the mathematical basis of an earth-centered cosmos—Ptolemy defends the study of the heavens on the grounds that it promotes the development of moral and religious qualities:

> With regard to virtuous conduct in practical actions and character, this science [mathematical astronomy] above all things, could make men see clearly; from the constancy, order, symmetry and calm which are associated with the divine, it makes its followers lovers of this divine beauty, accustoming them and reforming their natures, as it were to a spiritual state.[53]

The ultimate end of astronomy, as Ptolemy understands it, is moral and spiritual formation. Ptolemy's moral astronomy is thus wholly in keeping with the tradition established by Plato's *Timaeus*, in which the goals of the philosophical life are intimately related to the study of the cosmos.

What I hope to have shown up to this point is that the classical Greek engagement with nature, while often touted as an ancestor to modern science, was so imbued with theological and moral elements that its relationship to "science" as we now understand it is at best complicated. It is not just that astronomy and natural philosophy had some additional ethical elements that were largely peripheral and have now fallen by the wayside. It is rather that the study of nature was given a role in a broader philosophical enterprise that had moral goals and, quite often, theological presuppositions. Unlike anything in the modern sciences, the study of physics or natural philosophy was an exercise directed toward the transformation of the self. To claim that our science was born in ancient Greece is to overlook what for ancient Greek philosophers was the main point of the exercise.

Having offered this overview of the issue of ancient "science," I now turn to the question of "religion" in antiquity, and in particular early Christian identity in relation to classical natural philosophy. I have argued that there is not a recognizable "science" in antiquity, and now will suggest that there is not a recognizable "religion" either—although as we will see, the appearance of Christianity brings something quite new, including some of the preconditions for the modern conception "religion" that will appear in the seventeenth century.

What Is Christianity?

Much has been written on the identity of early Christianity, and it is not my intention to offer a comprehensive account of it here. But I do want to pursue this question insofar as it is relevant to the story that I wish to tell about the relationship between Christianity and the study of nature, and about how Christianity in the early modern period became the model for a generic "religion." This truncated account, then, will seek to identify those misconceptions about ancient Christianity that are analogous in many respects to those that distort our understanding of ancient philosophy and science. The basic claim will be that our notion "religion" is not present in the first centuries of the Christian era, although the emergence of Christianity is nonetheless accompanied by a new understanding of religious identity that is the precondition for the emergence of the modern idea.

It is instructive to begin with a late second-century Christian text that helpfully points to a range of possibilities for early Christian identity. The *Epistle to Diognetus* was written by an unknown Christian apologist to a highly placed pagan. In spite of its anonymity and relative brevity, the work has been described as "the noblest of early Christian writings" and as deserving "to rank among the most brilliant and beautiful works of Christian Greek literature."[54] It represents the earliest substantial attempt to give an account of the phenomenon of Christianity to outsiders. Here are its opening lines:

> Since I see, most excellent Diognetus, that you are exceedingly anxious to understand the godliness [theosebeia] of the Christians, and that your enquiries respecting them are earnestly and carefully made, as to what God they trust and how they worship [thrēskeuein] Him. . . . You wonder,

too, why this new race [*genos*] or way of life [*epitēdeuma*] has appeared
on earth now and not earlier.[55]

I will not offer any in-depth analysis of this passage, but allude to it, first,
to reinforce the point that the category "religion" was not then available
as a way for the first Christians to identify themselves: the author speaks
not of a system of beliefs and practices but of godliness, modes of worship,
a new kind of race, and a way of life.[56] Second, I want to suggest that with
the inception of Christianity we see the appearance of something quite
new — not simply another species of some existing genus "religion," but
a whole new genus, albeit one that draws to some extent upon existing
models. That said, what we do see in this new way of categorizing a set of
human activities and commitments are key elements of what in the seventeenth century will become the modern idea "religion."

A few brief comments on the terminology will reinforce these two
points. *Theosebeia* is a word that appears only once in the New Testament.
In translations of this passage it has been variously rendered in English:
"piety," "godliness," "godly piety," "fear of God," and "worship of God."[57]
Thrēskeia we discussed briefly in the previous chapter. Including variants, it occurs only five times in the New Testament. In its first appearance, Saint Paul makes reference to his having been a Pharisee, belonging to the "strictest sect of our *religion*," as it is usually translated.[58] This is
perhaps the closest to our modern sense of religion, although the context
suggests that it is moral discipline that is being referred to. In the second occurrence, the relevant phrase is usually rendered the "worship" or
"worshipping" of angels, although both Wycliffe and the Douay-Rheims
translations speak of the "religion" of angels.[59] By this they mean worship
directed toward angels. The other three references occur in the Epistle of
James, where it is typically rendered "religion."[60] As we have seen, in this
latter context the meaning is that true cult or worship that is acceptable
to God consists of charitable acts. The other two terms have no explicitly
religious connotations. *Genos*, from which our English words "genus," "generic," and "genetic" derive, refers predominantly to "race" or "kinds," but
also to birth, countrymen, descendents, nations.[61] *Epitēdeuma*, which does

not appear in the Greek New Testament at all, means something like "practice," "occupation," "pursuit," "training," or, as rendered in this passage, a
"way of life."[62] We might note, in passing, that "Christianity" (christianis-
mos) does not appear here at all. Moreover, although the word "Christian"
(christianos) is used here (and occurs three times in the New Testament), it
was an outsider's term for the early Christian movement, best understood
in the light of an imperial context in which the issue was the potentially
competing political allegiances of the Christians.[63]

None of these terms is suggestive of commitment to a system of propositions. Equally significant is the fact that the author of this passage, in
using four different descriptors in the space of a few sentences, is visibly
struggling to set out a conception of the Christian life that has no exact
precedents. In terms of what is new in this still-coalescing formulation, it
partly turns on its universal claims, encapsulated in the Pauline formula:
"neither Jew nor Greek, slave nor free, male nor female."[64] Christianity
thus conceived transcends the particularities of ethnicity, social status,
and gender. This conceptualization of a religious way of life that floats free
from specific cultural contexts is quite new. The Christian concept of "a
new race," then, is of a new kind of race, one that is open to all.[65] It is also
important that Christianity is not anchored to a particular geographical
space. Again this sentiment may be found in the New Testament where
in John's Gospel Jesus declares to the Samaritan woman that true worship
will take place not in a temple cult located in Jerusalem or on Mount Geri-
zim, but in the hearts of believers who "worship in spirit and in truth."[66] In
this new form of religiosity, cultic practice is interiorized so that the wor-
shipper is not bound to a particular sacred place. Neither, in spite of the
centrality of the historical events of Christ's life, is Christianity limited to
a particular time, since it was argued that in some form, Christianity had
existed since the beginning of time.

The coming into existence of this new identity has sometimes been referred to as the "disembedding" of religion—a process through which religion is separated out from the cultural forms with which it had been amal-
gamated. This is an entirely appropriate way of speaking, provided that it
is understood that religion is not some essential feature of human existence that lay dormant and awaiting this historical moment to emerge.[67]

Rather, this new understanding of *religio* was constructed by the first Christians, although, as we will see shortly, they themselves could sometimes write as if there were an ideal and eternal religion that received its definitive expression with the inception of Christianity.

If Christian identity was novel in these respects, there were also significant continuities. Saint Paul had identified the closest existing models, albeit not in the most enthusiastic terms, in his observation that the Christian Gospel was "a stumbling block to Jews and folly to the Greeks" and again in the double negation of "neither Jew nor Greek."[68] Note again that the categories to which Christianity is compared are closer to what we would call ethnic or cultural categories than religious ones—Jews and Greeks, not Jewish religion and Greek religion. But while this oppositional motif is conspicuous in the New Testament and in the writings of the Church Fathers, the first Christians also adopted positive understandings of their relation to their cultural milieu, the two most obvious models being that of the "New Israel" and "the true philosophy" (or philosophy as a preparation for the Gospel—*praeparatio evangelica*).[69] These designations reflect both continuity and difference, and we will briefly consider them in turn. What I hope is beginning to become apparent is that the first Christians do not represent themselves as a new "religion" and they do not confront pagan (natural) philosophy as a "science" but rather as an element of a competing spiritual practice. To some degree, both "Jewishness" and Greek wisdom will be understood as offering different and deficient versions of the spiritual life and the means to realize it. This will lead to the idea of "Judaism," "Hellenism," and Christianity as rival forms of *religio*.

When we come to speak of the relations between Christians and Jews in the first centuries of the Christian era we must immediately confront the fact that at this time Jewish identity was itself undergoing a profound transformation. Arguably, these changes would leave an indelible mark on the trajectory of Western culture.[70] Two elements of these transformations are of particular importance. The first of these we have already alluded to and concerns the mode of worship of the Jewish people. A key moment of transition was the destruction of the second Temple by the Roman general Titus in AD 70. This catastrophic event brought an end to the central ritual of first-century Jewish culture. As scholar of religion Guy

Stroumsa has observed, with the cessation of the Temple cult Jews came to experience sacrifice internally in terms of metaphor and myth. This transition contributed in a crucial way to what I have called the "interiorization" of religion.[71] As we have already seen, this change is paralleled in the passage on worship in John's Gospel, which was presumably written after the destruction of the Temple. Worship is now an internal act. This idea will be rehearsed 1,200 years later in Aquinas's contention that the primary acts of *religio* are internal ones.

Second, as a religion with a sacred book, Judaism contributes to "the emergence of a textual culture" and bequeaths to the fledgling Christian faith the idea of the divine word and a central text. While Judaism itself was moving away from the written mode to the oral, Christianity, aided by the new medium of the codex (which had begun to replace the more cumbersome scroll), placed a sacred text at the center of its practices and theological reflections.[72] This prioritization of texts, together with the move toward a new interiority, is evident in changing reading practices that see a shift from public recitation to private meditative reading. These developments are evident in Augustine's autobiographical reflections, and in the rise of the practice of silent reading.[73] This textual orientation will be crucial for the Christian approach to nature, for the natural world, too, will be conceived along the lines of a book, with the metaphor of the "book of nature" offering a new model of the meaning and purpose of the created world. Again, this is quite new, and will be taken up in some detail in the next chapter.

Finally, when considering the way in which Judaism influenced early Christian identity it is important to bear in mind that the changes that "Judaism" undergoes are themselves determined partly by the Christianization of the Roman Empire. That is to say that not only is Christianity influenced by Judaism, but that the shape that Judaism begins to assume in the third and fourth centuries owes something to a burgeoning Christianity. Judaism becomes a kind of religion that is akin in certain respects to Christianity.[74]

Folly to the Greeks?

Turning to the relation between Christian communities and classical culture—again an enormous topic in its own right—it can be said that the philosophy of pagan antiquity also contributed to a model of Christian identity, specifically through the idea of Christianity as the realization of the unfulfilled goals of pagan philosophy. Often the relationship between "Christianity" and "classical culture"—understood in these monolithic terms—is envisaged along the lines of Karl Popper's suggestion of an "intolerant Christianity" suppressing the philosophical and scientific achievements of the Greeks. This purported clash of cultures in the first centuries of the Common Era thus plays a role in the larger narrative of an ongoing conflict between science and religion. Undoubtedly, agonistic motifs are present in the writings of pagans and Christians alike. Saint Paul spoke dismissively of "the wisdom of the world" and of the perils of "philosophy and vain deceit" (I Cor. 3:19; Col. 2:8). North African Christian writer Tertullian (AD ca. 160–ca. 225) followed suit, famously declaring pagan philosophy to be the parent of heresy, and posing the redundant question: "What has Athens to do with Jerusalem?" Not surprisingly, some pagan philosophers responded in kind. The second-century Greek philosopher Celsus unflatteringly described Christians as "ignorant and bucolic yokels" who "believe without rational thought."[75] This assessment was subsequently endorsed by Porphyry, whose "Against the Christians" catalogs similar complaints. The Christians, he charges, "can establish nothing by way of demonstration, but hold to an unreasoning faith."[76]

These claims and counterclaims are couched in terms that suggest competing systems of knowledge. But there are features of these disputes that point in a slightly different direction. Porphyry's attack on the Alexandrian Church Father Origen (AD 185–253) is particularly revealing. Describing Origen's education at the hands of the Platonist philosopher Ammonius Saccas, Porphyry writes that Origen "derived much benefit from his teacher in the knowledge of the sciences; but as to the correct choice of life, he pursued a course opposite to his."[77] The point of difference between the pagan Ammonius and Christian Origen was thus not to do with their knowledge of the sciences, but their chosen way of life. The first his-

torian of the Christian Church, Eusebius of Caesarea (ca. AD 260–ca. 341),
makes a similar point, this time in relation to the Christian adoption of
Jewish teachings: "though gladly accepting their Scriptures, we decline to
follow their mode of life." Christianity, he goes on to say, "is neither Hellenism nor Judaism, but a new and true kind of divine philosophy."[78] The
relevant distinguishing feature of Christianity is related to neither the sciences nor scriptures, but to a mode of life. And his mode of life is understood as a new kind of philosophy.

This idea—that Christianity was the one true philosophy—is best
understood in light of the claim, made earlier in this chapter, that classical philosophy itself is primarily concerned with moral and spiritual
formation, and not dialectic and doctrine. Both the Church Fathers and
medieval scholastics perceived this more keenly than many modern commentators. Augustine expressed it this way:

> First let me tell you in general that there is one overriding concern com-
> mon to all philosophers, and that in this common concern they divided
> up into five different sets of special opinions. In common, all philosophers
> strove by dedication, investigation, discussion, by their way of life, to lay
> hold of the blessed life (*beata vita*). This was their one reason for philoso-
> phizing; but I rather think that the philosophers also have this in common
> with us. . . . Therefore the urge for the blessed life is common to philoso-
> phers and Christians.[79]

In the thirteenth century Thomas Aquinas would subsequently point out
that "the philosophers of the past had sought an 'ultimate felicity,'" but
that this had sadly eluded them: "since Aristotle saw that there is no other
knowledge for man in this life than through the speculative sciences, he
maintained that man does not achieve perfect felicity, but only a limited
kind." For this reason, he concludes, the philosophical endeavors of Aristotle, and the brilliant pagans that came after him, invariably ended in
anguish (*angustia*).[80]

The apparent ambivalence of the Church Fathers toward philosophy can
thus be accounted for in terms of their affirmation of the ends of philosophy, but skepticism about the means. As is well known, a number of Chris-

7

tian writers gave cautiously positive assessments of the classical quest for wisdom. Justin Martyr (AD 103-65) was to describe philosophy as "the greatest possession, and most honourable before God, to whom it leads us and alone commends us." The philosophers, he said, were "truly holy men."[81] Clement of Alexandria (ca. AD 150-ca. 220) characterized philosophy as "the work of divine providence" and "a handmaid to the Greeks." As the Law had been to the Hebrews, so, for the gentiles, philosophy was a "preparatory training" for the Christian Gospel.[82] Fellow Alexandrian Origen also thought that philosophy was a good preparation for Christianity and urged familiarity with the pagan sciences in order to assist Christian readers with their understanding of holy scripture.[83] Augustine himself frequently described Christianity as a "true philosophy" and also insisted on the advantages of pagan learning for biblical exegesis.[84]

In keeping with this qualified endorsement of the goals of pagan philosophy, a number of the Church Fathers commented on the received divisions of knowledge insofar as these were thought to represent stages of spiritual development. At its most general level, these stages were thought of as representing a progression that moved from the contemplation of visible things (natural philosophy) to things invisible (theological truths). Anaxagoras's formula according to which "manifest things are a vision of hidden things" thus matches the cadence of Saint Paul's contention that "since the creation of the world God's invisible qualities . . . have been clearly seen, being understood from what has been made."[85] Early Christian writers adopted versions of the classical divisions of knowledge, but related them to the goals of theological contemplation as understood within the Christian tradition. The Alexandrian Fathers Clement and Origen treated the divisions of knowledge as stages on the path to contemplation of God, and suggested that the Hebrew Bible had already anticipated these stages.

Origen, for example, saw in the training of philosophers in Plato's *Republic* an educative process in which the novice passed through successive stages of ethics and mathematics before proceeding to theological contemplation. For him, this philosophical training was a pale imitation of a parallel progression that could be accomplished through study of the wisdom literature of the Hebrew Bible. Origen proposed that the book of

Proverbs taught ethics; Ecclesiastes, natural philosophy; and the Song of Songs, theological contemplation. Thus the pattern of spiritual progress sought by the Greeks was already present in the Hebrew Bible, with the pagan divisions being plagiarized from this canonical original.[86] A number of Christian writers followed Origen in identifying these three books with three stages of spiritual growth.[87] Not surprisingly, these stages of contemplation were also incorporated into the ascetic practices of the monastics. In his primer on monastic exercises, John Cassian (ca. AD 360–430) suggested that these same three books represented three successive "renunciations." In the third stage, represented by the Song of Songs, the mind learns to move away "from every earthly affection . . . to things which are invisible, thanks to ceaseless meditation on divine realities and to spiritual contemplation [*theoria*]."[88] Aristotle's venerable idea that the most elevated human activity lay in contemplation (*theoria*) was thus subsumed by the Christian contemplative tradition, and his three speculative sciences found themselves mapped onto the three stages of mystical theology (*purgatio, illuminatio, unitio*).

Latin Christians undertook similar sublimations of these elements of Greek philosophy. Augustine suggested that the stages of learning proposed by the pagans could serve to draw the inquirer toward God. Following the Stoic (rather than the Aristotelian) classification of the sciences, Augustine explained that the object of physics is God as the ultimate cause of being; of logic, God as the criterion of thought; and of ethics, God as the rule of proper conduct.[89] Augustine also claimed that natural philosophy, moral philosophy, and logic were all contained in Christ's commandment to "Love God, and love neighbour"—a claim that would seem far-fetched, were it not for the moral orientation of pagan philosophy.[90] (Interestingly, in the seventeenth century, as we will see in chapter 5, the claim would again be made that natural philosophy was essentially to do with love of God and of neighbor.) There is also a progressive element in Augustine's understanding of the division of knowledge, because each stage prompted the development of distinct intellectual virtues. The inquirer begins with a knowledge (*scientia*) of earthly things, while moving toward a wisdom (*sapientia*) of heavenly things. Augustine explains that we go "through knowledge, towards wisdom." *Scientia*, the cultivation of which is promoted by natural philosophy, is thus a way station in the progression toward *sapientia*, the virtue or state of mind associated with theology. Even the most eminent of the heathen philosophers, Augustine explains, had failed to progress much beyond earthly things.[91] Hence, the danger of natural philosophy was not to do with its content, but with the possibility that it might be regarded as an adequate and self-sufficient path to wisdom and that its adherents might become preoccupied with the lesser goods of the material world. Such concerns would later reemerge in the thirteenth-century condemnations of aspects of Aristotelianism.

If we now consider early Christian identity in relation to both the Jews and the Greeks, it can be seen that a dominant motif—that philosophy for the Greeks and the Law for the Jews were a preparation for the Gospel—involves a mapping of the notion of fulfillment onto their respective histories. Thus, the idea that the one true philosophy has existed since the creation had its counterpart in the idea of a true Christian religion being incipient in the world from the time of its infancy, as evidenced by the promise given in Genesis 3:15, the faith of Abraham, and the prescience of the Hebrew prophets. The prophets "lived according to Jesus Christ," insisted Ignatius of Antioch (b. ca. 50).[92] The second-century Assyrian writer Tatian contended that "the philosophy of the Christians is more ancient than that of the Greeks."[93] Eusebius wrote of "an ideal of godliness" that must have guided the lives of the biblical patriarchs. This was in fact Christianity, which Eusebius describes as "a third order" that was neither Hellenism nor Judaism, but something between the two. Christianity was thus "the most ancient organization for piety, and the most venerable philosophy, only in recent times codified as the law for all mankind."[94] On the same theme, Augustine insisted that the coming of Christ signified a change in neither God nor religion, since true religion had existed from the beginning of the human race. This religion included the biblical patriarchs from Adam to Moses and the people of Israel, but also took in "others now and then at other times and in other peoples."[95] These sentiments, which are by no means uncommon among the Church Fathers, represent a further instance of the disembedding of Christianity, in this case in relation to historical time. Christianity floats free in time and can be extended back to the very beginning of history, laying claim to a lineage that begins with Adam, and is transmitted through both Jewish patriarchs and noble pagans. This extension backward provides an answer to the question posed by Diognetus, broached at the beginning of this discussion—why this "new race" has appeared now and not earlier. In fact Christianity had always been present. Looking forward, this new mode of religiosity—a disembedded Christianity—will be a precondition for the idea of a generic religion that will crystallize in the early modern period.[96]

Beliefs, Creeds, and Doctrines

There has been little mention to this point of the role of belief and doctrinal commitment in early Christian self-understanding. Yet, on a common view, this is the place to begin an inquiry about what Christianity is and how it relates to competing systems of knowledge such as philosophy and "science." Christianity, on this view, is defined primarily by what it is that its adherents believe, and its relationship to philosophy is determined by the extent to which these beliefs are consistent with philosophical doctrines. By now it should be clear that this understanding arises out of the distinctively modern idea that Christianity is a religion, and that religions are constituted by beliefs and practices. Premodern Christians, innocent of this modern conception, were unable to conceptualize their commitments in this way, and the various expressions deployed in the *Epistle to Diognetus*—godliness, worship, race, way of life—seem to point in this direction. It should also be clear that ancient philosophy, too, was less doctrinal than we have tended to imagine. Yet, clearly, doctrines have played some role in both philosophy and Christianity, and particularly the latter. After all, canonical books have a content, and Christianity took its sacred texts seriously. Those sacred books themselves contain references to "sound words," "the healthy doctrine," "the deposit," "the word of life," and so on.[97] Moreover, the first centuries of Christianity saw the appearance of "rules of faith" and the subsequent distillation of doctrines in the form of creeds and symbols, from very early creedal statements to the more formulaic Nicene Creed (AD 381) and the Symbol of Chalcedon (AD 451). These look very much like sets of propositions that constitute the essence of Christian belief. The development of theology, as we have already noted, comes much later. But nonetheless, this distinctively Western form of intellectual activity predates the modern period and also seems to underscore the importance of doctrine and correct belief.

It is not possible to offer here a complete analysis of the place of commitment to doctrinal propositions in early Christianity, but to begin with we can say that in certain respects, the issue of the role of doctrine in ancient Christianity is related to our earlier discussions about the place of doctrine in ancient philosophy. My suggestion has been that ancient philosophy has often been misunderstood by modern commentators who, on account of their adherence to a relatively recent and epistemologically oriented idea of philosophy, have seen only philosophical doctrines where in fact there were also philosophical exercises and ways of life. But it does not follow from this that doctrines played *no* role in these ways of life. As we have seen, for Plato and the Stoics teachings about the nature of the cosmos provide the context for an understanding of how life is to be lived. While Seneca, for example, stressed the fact that philosophy was concerned with the ordering of one's life, rather than doctrine and dialectic, he also resisted the extreme form of this position according to which moral precepts alone were sufficient for the philosophical life.[98] For moral precepts to be most effective, he pointed out, they must be grounded in a particular view of the cosmos: "Precepts by themselves are weak and, so to speak, rootless. . . . It is the doctrines which will strengthen and support us in peace and calm, which will include simultaneously the whole of the life and the universe in its completeness."[99] Doctrines thus provide a context for the justification of moral prescriptions, underscoring the fact that for the Stoics, natural philosophy—the attempt to understand "the universe in its completeness"—promotes moral edification. Doctrines inform the moral content of philosophy while not representing ends in themselves.

As well as offering a rational grounding for prescriptions relating to the good life, doctrinal formulas could also have a more direct therapeutic effect. This is most conspicuous in the link between philosophical and Christian understandings of catechesis. The expression "catechism" was used for the canonical doctrines of Epicureanism, and the rehearsal of the Epicurean creed was aimed at inculcating the appropriate attitude to life. "One must laugh and at the same time philosophize and look after his own affairs," wrote Epicurus, "and never cease to utter the words of the correct philosophy."[100] Philosophizing and the affairs of life were thus to be shaped by the mental rehearsal of the tenets of Epicureanism.

For the Church Fathers, catechesis and doctrinal understanding were similarly pursued in order to serve the ends of the religious life. One of the chief functions of the recitation of the Christian creed was thus to shape the direction of the lives of the faithful. "Say the creed daily," Augustine urged his catechumens, "when you rise, when you compose yourself to sleep, repeat your creed, render it to the Lord, remind yourself of it, be not irked to say it over."[101] Elsewhere he hints that the private recitation of the creed was an exercise in self-examination, echoing the sentiment of the Delphic maxim "know thyself": "look at yourself; treat your creed as your own personal mirror."[102] The formative function of the creed and its role as a starting point in the Christian pilgrimage are set out in the closing lines of his short *Treatise on the Faith and the Creed:*

> And these few words are known to the faithful, to the end that in believing they may be made subject to God; that being made subject, they may rightly live; that in rightly living, they may make the heart pure; that with the heart made pure, they may understand that which they believe.[103]

The creed is thus an introduction to Christian life, and believing is concerned with subjection to God and right living. These, in turn, will bring a full understanding of Christian faith. These priorities are reflected in one of Augustine's favorite sayings: "unless you believe, you will not understand."[104] His sermon on this dictum opens with the telling observation that "the starting point of a good life . . . is right faith."[105]

Consider also the role given to words in the philosophical therapies of the ancients. The basic premise of ancient rhetoric was that words, in addition to their referential and cognitive functions, have an affective dimension that is capable of moving and remediating the soul. In his dialogue *Charmides* Plato proposed that "the cure of the soul . . . has to be effected by the use of certain charms, and these charms are fair words."[106] Christian thinkers, who also regarded themselves as being concerned with the cure of souls, sought to use words in a similar fashion.[107] Traditionally, the cure of the soul had been addressed to the amendment of "false beliefs and evil habits."[108] While we now conceive of these as quite distinct shortcomings—located, respectively, in the spheres of knowledge and morality—in antiquity they were intimately connected. For this reason Seneca linked moral precepts with cosmological doctrines. Christian writers were similarly convinced of the inseparability of moral and epistemological error. The fall of Adam was thus thought to have precipitated the human race into a state of ignorance. On this understanding, not only did the original moral failings of humanity lead to false belief and proneness to cognitive error, but they continued to do so.[109]

The Church Fathers and their medieval successors thus routinely linked immorality and heresy. It was for this reason that Clement offered the otherwise puzzling remark that "the ideas entertained of God by wicked men must be bad, and those of good men most excellent."[110] Upright living and right belief go together. Heretical movements were rarely censured purely for the fact that they promoted false beliefs. The threat of heresy was moral and social. Constantine the Great (272–337), the first Roman emperor to adopt Christianity, was particularly concerned to ensure right worship in the empire for this reason. In a letter to the North African bishops, dispatched in the year 330, he speaks of "heretics and schismatics, who, deserting good and following after evil, do the things that are displeasing to God, are proved to cling to the devil, who is their father." So great is their "wicked and shameless perversity," he goes on to suggest, that "they might even break out into tumults, and stir up men like themselves at their crowded meetings, and thus a state of sedition might be produced, which could not be allayed."[111] (Interestingly, in this instance Constantine did not invoke the historical prerogative of Roman emperors to enforce proper religious worship, but concluded his letter with a counsel of patience and tolerance.) These associations of heterodox belief with improper worship, immorality, disloyalty, and sedition give a strong indication of the fact that religious belief was not a discrete variable of some notion "religion." Beliefs as we understand them were embedded in, and hence inextricable from, social and political realities. Returning to the homiletic practices of the Fathers, we can conclude that the preaching was less concerned with the communication of doctrine to the faithful than with the attempt to use a form of words to direct their spiritual growth and shape the boundaries of a moral community.

Another way of shedding light on the nexus between belief and practice is to consider the relevant terminology. As is the case with a number of the words that we are accustomed to using in a particular way—"religion," "theology," "science"—the vocabulary of creedal belief is particularly instructive. "Creed" derives from the Latin *credo,* a translation of the Greek *pisteuō.* These terms denoted something both more and less than the giving of assent to propositions.[112] *Pistis* thus originally meant "confidence" or "trust" between persons. In the New Testament, where the term takes on a more specific religious meaning, it has a cognitive component but retains the basic idea of trust, along with connotations of obedience, hope, and faithfulness. The Latin uses both a noun (*fides*) and a verb (*credere*—to believe; *credo*—I believe) to capture the meaning of the Greek, and this is paralleled in the English use of the noun "faith" and the verb "to believe" to render *pistis* and its cognates. Originally both the Latin and English retained the noncognitive connotations of the Greek. Thus, *credo* meant *"to trust to or confide in a person or thing, to have confidence in, to trust."*[113] When Seneca observes that "the first way to worship the gods is to believe in [*credere in*] the gods" he is not suggesting that believing *in the existence* of the gods is a form of worship. Rather, he means to say that the beginning of worship consists in placing one's trust in the gods. A sufficient condition for worship, he goes on to say, is imitation of the gods.[114] Augustine sets out a similar view, proposing that believing in God (*credere in*) means "to love Him, by believing to esteem highly, by believing to go into Him and be incorporated into His members." Elsewhere he contends that believing in God "is a great deal more than believing what God says": it is "to cling by faith to God."[115] The English term "believe" captured this sense nicely, and in its medieval and early modern usages meant simply "to trust, rely upon, or have confidence in a person."[116] Until well into the Middle Ages, then, the declaration "I believe" was neither an assertion of the existence of some being nor the lending of assent to propositional truths, but was primarily an expression of trust between persons. As Faustus of Riez (b. ca. AD 405) expressed it: "To believe in God is to seek him in faith, to hope piously in him, and to pass into him by a movement of choice. When I say that I believe in him, I confess him, offer him worship, adore him, give myself over to him wholly and transfer to him all my affection."[117]

The contraction of the scope of "I believe," and the assumption of its more familiar modern meaning—"to believe in the actual existence of some person or thing"—begins in the seventeenth century.[118] As we shall see in chapter 4, "believe in," from this time onward, is less a personal commitment than a theoretical judgment. It becomes a matter of disinterested intellectual assent to putative facts and, increasingly from the nineteenth century, to doubtful or false claims. It is no coincidence that these transitions are accompanied by the growth of the ideal of scientific objectivity and the gradual disappearance of moral and formative elements from natural philosophical discourse. For the moment, however, my suggestion is that "belief" is not a historically stable concept in the West, and thus it provides a poor foundation for a notion like "religion," for which is claimed both transhistorical and cross-cultural applicability. Indeed, if social anthropologist Rodney Needham is correct, the absence of our contemporary idea of belief is something that the premodern West shares with most other cultures. Needham points out that "there are numerous linguistic traditions which make no provision for the expression of belief and which do not recognise such a condition in their psychological assessments." "Belief" in our modern Western sense, he concludes, "is a relatively modern linguistic invention, and it does not correspond, under any aspect, to a real, constant, and distinct source of the self."[119]

Similar considerations apply in the case of "doctrine," which is often identified as the propositional substance of belief: doctrine is what is believed. If the analysis of belief set out above is correct, then this understanding of doctrine cannot be right. But if doctrine is not the primary object of belief, what is it? Again, attending to the history of the term is illuminating. In antiquity, *doctrina* meant "teaching"—literally, the activity of a doctor—and "the habit produced by instruction," in addition to referring to the knowledge imparted by teaching.[120] *Doctrina* is thus an activity or a process of training and habituation. Both of these understandings are consistent with the general point that Christianity was understood more as a way of life than a body of doctrines. Moreover they will also correlate with the notion of theology as an intellectual habit, as briefly noted in the

previous chapter. As for the subject matter of *doctrina*—its cognitive component, if you will—this was then understood to be scripture itself, rather
than "doctrines" in the sense of systematically arranged and logically related theological tenets. To take the most obvious example, Augustine's *De
doctrina Christiana* (On Christian Teaching) was devoted to the interpretation of scripture, and not to systematic theology.[121] Even in the thirteenth
century Bonaventure and Aquinas could both say that theology is nothing
other than sacred scripture.[122]

As we have already noted, an emphasis on the reading of texts was a
distinctive feature of Christianity. Here we can identify a significant point
of difference with philosophy and the status of philosophical "teachings."
For a start, philosophers had typically privileged the spoken word over
the written medium.[123] More importantly, one mark of the deficiency of
pagan philosophy, at least according to its Christian critics, was the plurality of teachings offered by the various philosophical schools. Augustine had made much of the fact that the Roman scholar Marcus Terentius
Varro (116–27 BC) identified a possible 288 philosophical sects. To be sure,
not all of these had been instantiated, but it followed, as Augustine mildly
expressed it, that "philosophers have expressed a great variety of diverse
opinions."[124] The Roman philosopher Cicero had already given a candid assessment of the enterprise, confessing that there is nothing so absurd that
some philosopher has not said it.[125] For Christian apologists the discord of
philosophy was routinely contrasted with the unity of scripture. From the
Christian perspective, this diversity of human opinion demonstrated the
fallibility of reason and necessity of revelation. Canonical scripture thus
served as a measure or rule against which teaching could be gauged. The
truth of scripture was clearly important, for upon its veracity rested the
truth of the Christian way of life. As Augustine put it, "not even love itself,
which is the end of the commandment and the fulfilling of the law, can be
rightly exercised unless the objects of love are true and not false."[126] For
this reason it became important to spell out rules for the interpretation
of scripture: hence works such as Augustine's *De doctrina Christiana*. The
creeds, too, in addition to their role in Christian formation, were understood as providing a framework for the understanding of scripture, and as
guides to the bounds of legitimate reading.

While it was important that the narratives of scripture be regarded as
true, the study of the Bible for the Church Fathers and their medieval successors was primarily a matter of becoming acquainted with the moral
demands of scripture and, by enacting those in daily life, growing in the
likeness of their divine author. Augustine wrote that the criterion of any
interpretation of scripture was whether it had a tendency "to build up
two-fold love of God and our neighbour" or to promote "the kingdom of
charity." This priority was in keeping with the Pauline maxim that knowledge puffs up, while charity edifies (I Cor. 8:1). Indeed, once the theological virtues have been formed in the individual, doctrine becomes unnecessary: "Thus, a man who is resting upon faith and hope and love, and who
keeps a firm hold on these does not need the scriptures except for the purpose of instructing others."[127] Gregory the Great (ca. AD 540–604) wrote
in a similar vein that in reading scripture "we should transform within
ourselves what we read, so that when the mind is moved by hearing, our
way of life puts into practice what it has heard."[128] The reading of scripture was not primarily a matter of learning the content of "the Christian
religion" as we might now understand it, but was a form of spiritual exercises akin in certain respects to those of the philosophers. As Brian Stock
expresses it, "Scripture offers the reader—either the private reader or the
audience at a reading—a privileged medium, through which God's will,
framed in narrative, can be internalized and directed outwards as ethi-
cally informed action."[129] The encounter with scripture was thus the occasion for self-transformation. As in the case of ancient philosophy, doctrine
in the sense of "teaching" plays an important role in patristic and medieval Christianity.[130] But just as ancient philosophies are not to be thought
of simply as assemblages of argumentative techniques and tenets, Christianity is not yet understood as *a* religion, constituted by its doctrinal
content.

:::

I hope to have identified some reasons for thinking that neither of the
categories "science" or "religion" works very well for this period, and that
if we attempt to apply them in some rigid fashion, the result is likely to be
a distorted understanding.[131] Even if we were to shoehorn particular activi-

ties of the ancient world into our categories "science" and "religion" as best
we could, we would still not get anything like the Karl Popper narrative in
which the natural philosophy of the Greeks represents a reaction against
theological interpretations of nature, and in which Christianity brings a
premature end to a fledgling science by reinstating them.[132] Rather, what
we see is a natural philosophy that is always pursued with moral and religious ends in mind. We know this because the relevant historical actors
state it unambiguously. Christianity can thus affirm the ends of pagan
natural philosophy, while rejecting the means. This realization enables us
to tell a rather different story to commonly received views. Three specific
aspects of this alternative narrative are worth briefly highlighting.

First, the idea that natural philosophy or science was subordinated to
theology by the Church Fathers and, after them, the scholastics, is quite
mistaken. As we have seen, it was Aristotle who places theology, albeit in
his own sense of the term, at the apex of the speculative sciences, and the
Stoic Chrysippus who contends that theology is the fulfillment of philosophy.[133] When in the Middle Ages Thomas Aquinas subsequently wonders
whether theology is "the ruler or mistress" of the other sciences, he does
so because the question is prompted by his reading of Aristotle. The scholastics were themselves not even agreed on whether theology was a science, and hence there was no consensus about its status as their queen.
The standing of "theology" in the ancient philosophical schools serves as
a reminder of the fact that philosophy, including natural philosophy, was
not then divorced from religious concerns.

Second, to a degree, Christianity *is* responsible for a diminution of the
scope of natural philosophy, but in the sense that it begins to take over
some of the moral and religious goals that were once intrinsic to natural philosophy. Further developments in the modern period will complete
this process, but with the inception of Christianity it has already begun. If
we consider again the uses of natural philosophy set out by Simplicius—
from the most mundane to the more elevated and spiritual—we already
have a good idea of which ones Christianity will consider itself responsible
for. This, then, is already a step in the direction of that demoralized and
detheologized version of natural philosophy that we designate "science."
What is left for science, thus conceived, are the residues of a natural phi-

losophy evacuated of those moral and spiritual components that had once lain at its heart. These amount to the mundane technologies of medicine and mechanics, and the production of engines of war.

Third, and related to this, it is not yet possible to set up an opposition between naturalistic and religious accounts of the cosmos. While some popular histories of science speak in terms of Greek science *versus* Christian faith, in fact, to some degree we might say that "theological" concerns could drive the agenda of naturalism.[134] The early Christians, for example, were on the whole more skeptical than their pagan counterparts about astrology, anthropomorphic deities, the world-soul, and the divinity of the heavens. To take just one these instances, Christians, Jews, and Epicureans were unique in the ancient world in not believing in the divinity of celestial bodies, and in the case of Jews and Christians this skepticism was motivated by a theological worldview. Educated pagans such as Celsus ridiculed Jews and Christians for their impiety in this matter. Partly on this account, Christians were often labeled as atheists and lumped together with the Epicureans, from whom, not surprisingly, they sought to distance themselves.[135] This reputation persisted into late antiquity with the pagan Simplicius, from whom we heard earlier, being horrified at the blasphemy of the Christian philosopher Philoponus (ca. 490–570), who denied divinity to heavenly bodies.[136] In fact Philoponus believed that the motion of the heavens was to be explained by a "motive force" imparted by God at the moment of creation. In his view, all natural motion was thus imparted, and he may, on this account, be credited with having supposed a unified theory of dynamics. His conception of impetus subsequently influenced Galileo.[137] But setting aside for now the apparent prescience of Philoponus, we can say that the Christian cosmos is not inhabited by deities. Yet, as a divine creation it does bear deep theological significance. Medieval Christians were to reinvest this depopulated world not with divinities but with divine meanings. In keeping with the textual orientation of Christianity they were to hold that nature, like scripture, was "God's book." The rise and fall of this powerful metaphor—the book of nature—along with its implications for our modern conceptions of science and religion, will be considered in detail in the next chapter. For now, however, consider that Christian critique of pagan philosophy is often interpreted as evidence of bias against "science." In fact much of that critique was directed against astrology, divination, the worship of deified heroes, and belief in the divinity of the celestial bodies, which is to say, against "superstition."[138] To speak of this as a conflict between science and religion is not merely to use inapplicable modern categories—it also involves a fundamental misunderstanding of the complicated way in which the cosmos was understood in Western antiquity.

[3]

Signs and Causes

According to the Gospel, the beginning of the ways of God is in
His work, so that the race of men might learn by Him to follow the
ways of the Lord and to perform the works of God.
—Ambrose of Milan, *Hexameron*[1]

By means of what is material and temporary we may lay hold upon
that which is spiritual and eternal.
—Augustine, *De doctrina christiana*[2]

Nor is it strange that philosophers through the whole of
speculative philosophy have diffused ethical principles, because
they knew that these related to man's salvation.
—Roger Bacon, *Opus Majus*[3]

In a recent collection of essays, addressed in part to the relations between
science and religion, the philosopher Thomas Nagel observes that contemporary analytic philosophers often fail to acknowledge "the significant element of yearning for cosmic reconciliation that has been part of
the philosophical impulse from the beginning." The example that Nagel
chooses to illustrate his point is Plato, who exhibited what he refers to as
"a profound religious temperament." This ancient Greek philosopher "was

clearly concerned not only with his soul, but also with his relation to the universe at the deepest level," and he "hoped for some form of redemption from philosophy."[4] Nagel's suggestion that Plato was embarked upon a "religious" quest, and that knowledge of the universe was integral to that quest, comports with the understanding of the goals of ancient philosophy set out in the previous chapter. Here, natural philosophy—what to us looks very much like science—is pursued as part of a comprehensive attempt to understand one's place in the world, and to live rightly within it.

As we have seen, the first Christian thinkers shared this religious temperament, but their understanding of our place in the cosmos was informed by an idea of creation that was in some tension with the predominant classical models. The idea that the world had been created out of nothing (*ex nihilo*) by a benevolent Deity contrasts with the Aristotelian doctrine of the eternity of the world, with Platonic and Gnostic teachings about the inferiority of the material world, and with the Neoplatonic idea of the world as an emanation from the divine. Christians, moreover, had the additional resource of scripture to assist with their version of the quest for redemption. For them, the normative qualities of the cosmos—that is to say, the manner in which it was thought to provide resources for the living of a fulfilled life—not only were to be understood in light of the idea of divine creation, but also had to be consonant with the content of the canonical texts of Christianity. For this reason the Church Fathers were to speak of two linked modes of divine communication—one in the book of scripture, and one in the book of nature. While the natural world was distinct from God, and hence nondivine, in various ways the creatures were thought to bear mute testimony to their divine origins. In a certain sense, the world was the bearer of the divine image, although, owing to the fallen condition of the cosmos and its human tenants, that image was faint and difficult to discern. But with the guidance provided by scripture the language of the book of nature could be understood. Reading scripture and nature together become an integral part of medieval contemplative practice.

The intelligibility of nature lay not only in the realm of meaning, however. As a divine artifact, the world was also causally related to God—as pagan thinkers, albeit in different ways, had recognized. In the Christian understanding of this relation, God was not only the efficient cause of things—the original cause that had brought them all into being—but was understood to be the source of the matter and form of the creatures, and the end or final cause to which all things were drawn. The study of nature, on this understanding, pointed to God as the causal power that ultimately underpinned the existence of all things.

For medieval thinkers these two conceptions of the divine relation to the world offered ways of ascending from the visible, mutable, and material things of nature to an intelligible spiritual realm that lay beyond. As exercises in the contemplation of nature, they also served to shape the person and to prepare the mind for the encounter with more elevated theological truths. From the thirteenth century onward the causal order will begin to predominate, as Aristotelian natural philosophy takes hold in the medieval universities. But both symbolic and causal understandings of the order of nature will operate in tandem until they are eclipsed by the new sciences of nature of the seventeenth century.

The Order of Meanings

The premises of the metaphor of the book of nature may be found in scripture itself. According to the book of Genesis, God placed lights in the heavens "for signs and for seasons" (Gen. 1:14). The heavenly lights thus served not only a practical purpose but for later interpreters acted as "signs" of eternal verities. Psalm 19 goes further, suggesting that the heavens communicate through a silent, wordless discourse: "The heavens declare the glory of God, the skies proclaim the work of his hands. Day after day they pour forth speech" (Ps. 19:1–2). In the New Testament, the key passage is Romans 1:18–20, which, as we have seen, alludes to the fact that the invisible things of God can be known through the visible creation. This passage provided a crucial link to the Greek divisions of the speculative sciences, in which the inquirer proceeds from the mutable, temporal objects of this world to the invisible objects of metaphysics and theology. This passage also afforded considerable scope for patristic and medieval writers to elaborate on how the contemplation of the creatures ought properly to lead to the contemplation of the Creator himself.[5]

Origen takes up these themes in his commentary on the *Song of Songs*—

the work most closely associated by patristic and medieval thinkers with the ascent of the mind beyond the realm of the material world. Here he suggests that "all the things in the visible category can be related to the invisible, the corporeal to incorporeal, and the manifest to those that are hidden." For Origen, there are hidden theological and moral meanings in all created things for, he argues, in much the same way that man bears an image and likeness to God, so all created things have a resemblance and likeness to heavenly things. Origen contends that these deeper layers of meaning are also to be found in the scriptures. While the superficial reader might rest content with the literal sense (which for Origen is equivalent to the visible category), more profound meanings are to be discerned in the deeper allegorical and moral layers that take us into the hidden depths. While not all patristic exegetes agreed with the specifics of this approach, and while we encounter multiple reading strategies among the Church Fathers, Origen is typical in so far as his whole interpretative endeavor is shaped by the Christian story of redemption that extends from the origins of the cosmos to the final consummation of all things.[6] This story is not only recounted in the words of scripture, but is built into the very fabric of the cosmos and is inscribed on every creature. Nature, like scripture, is a book upon which we may contemplate.

A number of the Church Fathers, following this logic, considered nature to be another canonical book. Athanasius (ca. 296–373) spoke of the way in which the creation, "as though in written characters" declares through its order and harmony the Lord and Creator. John Chrysostom (ca. 349–407) described the realm of visible things as a volume from which the unlearned may be educated in the things of God. Augustine chastised Faustus the Manichean for not looking upon the "book of nature" as the creation of God.[7] At times, this image could be strained to its metaphorical limits. Evagrius Ponticus (345–99), one of the most influential monastic writers of the patristic period, suggested that when we read the book of nature our own minds become books in which God inscribes words of wisdom and providence.[8] Although this conception of the order of nature is different from what we find in the classical writers, it still draws upon the idea of personal transformation as a key element of the engagement with nature. The human mind is itself rewritten as a consequence of its contemplation of the world.

In the patristic period and the Middle Ages the idea of the book of nature, and the question of how that "book" was to be read, were intimately linked to theories of the interpretation of that other book, scripture. Following Origen, a number of Church Fathers developed three- and fourfold schemes of biblical interpretation. The most common medieval classification distinguished four senses of scripture: literal or historical, tropological, anagogical, and allegorical.[9] The literal sense is more or less self-explanatory. The tropological sense referred to the moral application of the text—how it could be put into practice. The anagogical sense referred to the promises of scripture and the foretaste of heaven—what was to be hoped for. Finally, the allegorical sense places Christ in the center of history. John Cassian, one of the first to employ this scheme, offers an example of how this might work in practice, by explaining the different meanings of "Jerusalem." A biblical reference to this place would denote, in its literal sense, the city of the Jews; in its tropological sense, the human soul; in its anagogical sense, the heavenly city of God; and in its allegorical sense, the Church of Christ.[10]

For our purposes what is significant is that all of these exegetical schemes, whatever their terminology, posit a basic division between the literal sense and the higher senses. These higher senses were collectively known as "the spiritual sense" or simply just "allegory."[11] Augustine, in opening sections of *De doctrina christiana,* provided the definitive rationale for the allegorical interpretation of scripture. Here we encounter an influential theory of natural signs in which the interpretation of words is linked to the meaning of objects. Augustine explains that allegory is not simply a technique for reading written texts, but is to do with discerning the moral and theological meanings of natural things. The literal sense of scripture can be determined by linking words to the things to which they refer. The allegorical meanings, however, lie in the meanings of the objects. As part of his work of creation, then, God invested the creatures with theological meanings that can be "read" by the discerning reader.

Later medieval references to the book of nature tend to assume this

Augustinian framework. Hugh of Saint Victor (ca. 1096–1141) explains the metaphor in this way:

> For the whole sensible world is like a kind of book written by the finger of God—that is, created by divine power—and each particular creature is somewhat like a figure, not invented by human decision, but instituted by the divine will to manifest the invisible things of God's wisdom.

Every creature, on this understanding, has been designed to manifest some divine truth. Hugh explicitly invokes the Augustinian theory of signs, insisting that "in divine utterance not only words, but even things have a meaning." Our knowledge of this symbolic order is not purely theoretical, for it changes the mind through a process of habituation. "It is . . . a great source of virtue for the practiced mind," Hugh informs us, "to learn, bit by bit, first to change about in visible and transitory things so that afterwards it may leave them behind altogether."[12] The Franciscan philosopher and theologian Bonaventure (1221–74) was subsequently to agree that "all creatures are essentially a certain image and likeness of Eternal Wisdom." They are offered to us "so that proceeding from the sign to the thing signified, these minds of ours may be guided through the sensible objects they do perceive to the intelligible world they do not."[13]

The theory of signs found in *De doctrina christiana* is also rehearsed in the very first question of Thomas Aquinas's monumental *Summa theologiae*. Here Thomas considers the various schemes of allegorical interpretation before explaining that they all rest upon the assumption that things in the natural world, like words, are capable of signifying: "The author of Holy Writ is God, in whose power it is to signify His meaning, not by words only (as man also can do), but also by things themselves." He goes on to say that the senses of scripture "are not multiplied because one word signifies several things, but because the things signified by the words can be themselves types of other things."[14] Necessarily, then, any full account of divine self-disclosure will entail reference to both scripture and nature. For the early Middle Ages the intelligibility of nature lay primarily in its moral and theological meanings, rather than in sets of causal relations.[15]

Figure 2. Illustration of the pelican from Ashmole Bestiary.

A specific application of this theory of signs to the natural world was given in the medieval bestiaries, themselves elaborations on an earlier work known as the *Physiologus*. In the bestiaries we find animals, plants, and stones invested with theological and moral meanings, so that the reader, when encountering these natural objects, will be put in mind of their true significance. The description of the pelican set out in these works is perhaps one of the clearest examples of how the creatures were interpreted (see figure 2). The key element of the description in the *Physiologus*, related in all of the bestiaries, is as follows:

> David says in Psalm 101, "I am like the pelican in loneliness." . . . If the
> pelican brings forth young and the little ones grow, they take to strik-
> ing their parents in the face. The parents, however, hitting back kill their
> young ones and then, moved by compassion, they weep over them for
> three days, lamenting over those whom they killed. On the third day,

their mother strikes her side and spills her own blood over their dead
bodies . . . and the blood itself awakens them from death.[16]

This account is followed by commentaries on the various meanings of the
creature. The Aberdeen Bestiary (ca. 1200) explains that the pelican signifies Christ, Egypt, and the world. It symbolizes Christ because "it kills its
young with its beak as preaching the word of God converts the unbelievers. It weeps ceaselessly for its young, as Christ wept with pity when he
raised Lazarus. Thus after three days, it revives its young with its blood,
as Christ saves us, whom he has redeemed with his own blood."[17] Further
explanations are offered for why the pelican also signifies Egypt and the
world.

On account of this description the pelican became an enduring emblem
of Christ's atoning death, and is ubiquitous in medieval art, architecture,
and literature. In Oxford, it can still be seen perched precariously on the
pulpit of Christ Church Cathedral and above the famous sundial in the
Corpus Christi quad. It adorns the lectern of Norwich Cathedral, and it can
be seen in the painted glass of Bourges Cathedral and above the altar of
Saint Giles-in-the-Fields, London. In his famous Eucharistic hymn, *Adoro
te devote,* Thomas Aquinas speaks of the pelican; Dante refers to Christ as
"nostro pellicano" and, less happily, Shakespeare's Lear regretfully refers
to Regan and Goneril as "those pelican daughters."[18]

In addition to these signifying functions, creatures could serve more
directly as moral tutors. This was the "tropological" way of reading the
world, in which animals and plants offered patterns of behavior to emu-
late or avoid.[19] The classic example was the industry of the insects. The bee
was commended in scripture as a good worker, Ambrose of Milan (374–97)
pointed out, and we are thus enjoined "to follow the example of that tiny
bee and imitate her work." Ambrose's *Hexameron* is replete with instances
of such creaturely exemplars: birds were "examples for our own way of
life"; the fishes "constitute for us a pattern of the vices"; the example of the
vine serves "for the instruction of our lives."[20] The *Hexameron* of Basil the
Great (330–79) and the *Moralia in Job* of Gregory the Great (540–604) also
enumerate the manifold ways in which observation of the creatures can

yield important moral lessons. The whole world, Basil concluded, "is truly a training place for rational souls."[21]

In a period well before evolutionary theory offered an explanation of the sheer quantity and diversity of living things, it made sense to think of God as having provided various creatures for food, clothing, medicine, and other uses of life—including, in the case of less welcome creatures, as punishment for sin. But the vast array of living things exceeded by a significant margin all that could be encompassed within those categories. The question—why pelicans, and indeed why any number of living things—could be answered by referring to their moral and symbolic significance.[22] Some were to be understood literally, as it were; others in terms of the truths they represented or the morals that they taught. As Richard of Saint Victor (d. 1173) explains: the "visible works of the Creator have been created and disposed for this, that they may both serve the needs of the present life and also exhibit a shadow of future goods." It follows, he goes on to say, that "we ought to understand some things simply and seek nothing in them according to the mystical sense, while some things . . . are capable of representing something according to the mystical sense."[23] This is a rather different natural world from that which we currently inhabit, and it was also different from the world of the ancients. Understood in these moral and symbolic terms, it was a world that was not primarily to be explained causally, nor exploited materially, but to be read and meditated upon.

The Uses of the Creatures

The fact that the world was understood as providing for the material, moral, and spiritual needs of its human inhabitants meant that the Church Fathers adopted what in some respects was a "utilitarian" approach to nature. Ambrose claimed that "there is nothing without a purpose. . . . What you consider as useless has use for others."[24] Basil wrote that everything in the world had been created "to contribute to some useful end and to the great advantage of all beings."[25] These kinds of sentiments were by no means unprecedented. Aristotle had famously declared in the *Politics* "that nature has made all the animals for the sake of man," and the Stoics who came after him endorsed this position enthusiastically. Anthropocentric attitudes to nature are not to be solely attributed to the Judeo-Christian tradition, and neither are they unique to the West.[26] But there is something quite distinctive in the different kinds of uses of natural things attested to in patristic and medieval sources. The uses of things, as we have seen, were extended to the moral and spiritual realm. But beyond this, they were also understood within the context of the Christian Fall-Redemption narrative.

In *De doctrina christiana* Augustine not only provided the semiotic theory that underpinned the subsequent allegorical reading of the "two books," he also linked his theory of signs to an important distinction between use (*uti*) and enjoyment (*frui*).[27] Immediately following his discussion of signs and things, he goes on to suggest that the things God has made are to be used, and he places that use within a specific understanding of the earthly condition. The present life is compared to that of the wayfarer passing through a strange and beautiful land on the way to his rightful home:

> We have wandered far from God. If we wish to return to our Father's
> home, this world must be used, not enjoyed, that so the invisible things of
> God may be clearly seen, being understood by the things that are made, —
> that is, by means of what is material and temporary we may lay hold upon
> that which is spiritual and eternal.[28]

As part of God's creation, material things are, in an important sense, good—as Augustine maintains against the Manicheans—but they are good in a relative sense. As sojourners we must use the temporal and material things of the present world to direct our minds to the eternal and immaterial things of the next. The distinction between use and enjoyment is revisited in Augustine's notion of the order of love. In essence, because there is a hierarchy of goods, our loves must be ordered and directed accordingly.[29]

Aquinas will later take up these themes, making explicit the tension between the goodness of the creatures as God's creation and the necessity of using them as a means to discover God. In a discussion of the role of knowledge in human happiness, he declares that "man's beatitude consists, not in considering creatures, but in contemplating God." But it does not follow that we should have no interest in the creatures at all. Rather "man's beatitude does consist somewhat in the right use of creatures, and in well-ordered love of them: and this I say with regard to the beatitude of a wayfarer."[30] Aquinas subsequently invokes the Pauline reference to "the invisible things of this world" to explain that since "God's effects show us the way to the contemplation of God Himself . . . it follows that the contemplation of the divine effects also belongs to the contemplative life, inasmuch as man is guided thereby to the knowledge of God."[31]

The fallen condition of the world and its human inhabitants has an important bearing on all of this. According to the history of the first human beings set out in the opening chapters Genesis, Adam and Eve had been created with all the perfections necessary for a happy life, and had been placed in a paradise that would meet all their needs. Yet, in spite of this, they had disobeyed God and were cast out of Eden. As a consequence, our first parents and their progeny now suffer all the indignities of mortality, including a propensity toward evil and error. Equally important, the natural world found itself implicated in the rebellion of Adam and Eve, becoming wild and unmanageable. Our tendency to misidentify what should be used and what should be enjoyed, the disordering of our loves, and our difficulties in discerning the true meanings of the creatures all spring from sin. In his original and perfect state, Adam had been able directly to intuit the uses of the creatures and their spiritual significance. Following the Fall, however, the luminous transparency of the creatures had faded, and the human mind had become darkened. Bonaventure suggested that Adam's original dominion over the creatures in the Garden of Eden consisted in a mental mastery of what it was that they represented. In the state of innocence, Adam "possessed knowledge of created things and was raised through their representation to God and to his praise, reverence, and love." As a consequence of the Fall, this knowledge was lost, and the powers of the mind—sense, imagination, reason, understanding, intelligence, and moral discernment—"were distorted by sin." The recapturing of this lost mastery could be achieved only if the powers that had originally made it possible were "cleansed by righteousness, trained by learning, and perfected by wisdom."[32] All of this could also be related back to the "two books" metaphor. Bonaventure makes the connection this way:

> The whole world is a shadow, a way and a trace; a book with writing front and back (Ezekiel 2:9). Indeed, in every creature there is a refulgence of the divine exemplar, but mixed with darkness. . . . When the soul sees these things, it seems to it that it should go through them from shadow to light, from the way to the end, from the trace to the truth, from the book to veritable knowledge which is in God.[33]

Formal learning—including knowledge of the natural world—thus had a role to play in restoring to the mind some of its original powers and perfections.

In the same vein, Hugh of Saint Victor had already argued that the liberal arts are to be pursued in order to restore a divine likeness in the human mind—again, a likeness that had been lost as a consequence of the Fall.[34] The cognitive imposition of order on the creatures, in which we come to an understanding of their moral and allegorical significance, is actually a way of reestablishing a dominion over them that had been lost as a consequence of the Fall. In this way, the arts and sciences provide a means for the partial restoration of the prelapsarian perfection of the world. Hugh contended that the goal of the arts is "restoring our nature's integrity, or the relieving of those weaknesses to which our present life lies subject." Philosophy, he stresses, echoing both Cicero and Augustine, is a "cure" or a "remedy."[35] Similar motifs may be found in John Scotus Erigena, Michael Scot (ca. 1175–1253), and Vincent of Beauvais (d. 1264).[36]

Looking forward to the seventeenth century, a number of features of these medieval approaches to nature will reappear, albeit in a different register. The internalized understanding of the mastery of the creatures will be turned outward onto the visible world, and the redemptive exercise will be understood as the control and manipulation of material things in the most literal sense. But the reestablishment of dominion over nature will be retained as one of the chief goals of the new science. The Fall narrative will also occupy a central position, motivating scientific inquiry and shaping its methods. Finally, ideas about the usefulness of natural things will take on an entirely new complexion, when individuals such as John Calvin and Francis Bacon reconfigure the idea of utility and challenge the traditional priority given to the contemplative life.[37]

The Order of Causes

With the collapse of the Western Roman empire in the fourth and fifth centuries, the textual remains of classical learning had been preserved largely in encyclopedic digests, with the result that the early Middle Ages was heir to a rather "thin" version of traditional Greek learning. The twelfth century witnessed the rediscovery of a "thick" version of the classical tradition—mostly represented in the works of Aristotle—that had been preserved and elaborated in the Arab world and the Eastern empire. This rediscovery was accompanied by renewed urbanization and the rise of the medieval universities—Bologna (1150), Paris (ca. 1200), and Oxford (1220). Simplifying the matter somewhat, we may say that we now have a new site of intellectual activity, the medieval university, and a new philosophical conversation partner in the person of Aristotle. These new sites of learning, with their generally positive approach to Aristotelian thought, do not displace monasticism or Platonic Augustinianism, but they do offer a revised model of the nature and purpose of natural philosophy and of its relation to theology, and hence a different conception of what each entails.[38]

Part of the impact of Aristotelian thinking was to bring a revaluing of the material world in its own right and a reconsideration of how knowledge of the world related to knowledge of divine things. A key principle was Aristotle's insistence on the priority of sensory knowledge. Whereas the trope of the relation of the visible to the invisible had traditionally been read as endorsing the idea of an ascent from the material to the intelligible realm, it could also be interpreted more in an Aristotelian way. Thomas Aquinas, for example, seeks to preserve the idea of an ascent from the material to the intelligible, but this is now expressed in terms of the Aristotelian principle that there is nothing in the intellect that was not first in the senses.[39] From this principle Aquinas draws this conclusion: "Whatever is said of God and creatures is said according as there is some relation of the creature to God as its principle and cause, wherein all the perfections of things preexist excellently."[40] For this reason also, he says, we can affirm the metaphors and allegories of scripture, which seek to point to God by referring to material realities. But the key thing is that one of the primary ways in which we know God is as the ultimate *cause* of the creatures.[41]

In addition to their symbolic capacities, then, the creatures provide a way to understanding God as the causal source of their being, since to some degree effects resemble their causes. This is a presupposition of Aquinas's famous "analogy of being."[42] This assumption can justify the more formal study of the natural world as represented in Aristotle's writings on natural science—the *libri naturales*—that came to occupy an important place in the arts curriculum of the medieval universities. To take a single example, Thomas's teacher at the University of Paris, Albertus Magnus (d. 1280), commenting on Aristotle's *Parts of Animals,* notes that "from the knowledge of these vile animals we can ascend to the knowledge of the first cause just as we ascend from effect to cause."[43]

It is important to understand that God was not understood simply as the first in a series of efficient causes. Rather, for each natural effect, both God and the natural agent are involved. God's role in natural causation was understood in terms of both his ongoing conservation of the being of natural things and of his communication to them of a likeness of his own causal capacity. When events take place in nature, it is not that God plays some remote causal role simply by virtue of his having initiated a chain of causes in the past that eventually brings about a particular outcome. Neither is the present causal activity to be understood as partly attributable to God and partly to nature. Instead, natural events are "wholly done by both, according to a different way."[44] God's causal activity is analogous to natural causation, but cannot be unequivocally identified with it.

Sometimes, modern readers see this discussion of God's role as a first cause as depicting a particular kind of relation between theology and Aristotelian natural philosophy (or "religion" and "science" if you will). The systematic study of nature, exemplified in Aristotelian natural philosophy, plays the role of science, and in that role provides premises for a natural theology—that is to say, evidence of the existence and nature of God derived from the natural world alone. This would represent a significant breach with the symbolic approach. While this reading is not completely misplaced, it is only part of the story, and perhaps not the most important part. My suggestion is that the appropriators of Aristotle are less interested in using his "science" to support theological propositions than they are in seeing the whole exercise as promoting particular kinds of virtues, broadly understood. What we have here is a kind of mental training that prepares the mind for the acknowledgment of certain truths.

William of Auvergne (d. 1249), one of the first Parisian theologians to grapple with the newly translated works of the Greek philosophers, contended that the study of the universe leads to both "the exaltation of the creator and the perfection of our souls."[45] Albert the Great subsequently wrote of the generic benefits of study, observing that training (*exercitium*) "awakens a capacity, not only concerning its object, but of another object as well. A person who is capable of seeing the truth in one thing is disposed to seeing it in another."[46] As we have already established, "science," on this understanding, is a mental habit rather than a body of knowledge that offers premises for arguments. This understanding of science, moreover, applies across the board—to the science of nature (natural philosophy) and to the science of God (theology). Aquinas will make a similar point when he insists that understanding God from the creatures is not an exercise in what we would call "philosophy of religion," but is part of the life of contemplation: "Since, however, God's effects show us the way to the contemplation of God Himself, according to Rm. 1:20 . . . it follows that the contemplation of the divine effects also belongs to the contemplative life, inasmuch as man is guided thereby to the knowledge of God."[47]

It is important, then, to reduce neither natural philosophy ("science") nor *sacra doctrina* ("theology") to familiarity with an organized body of doctrines.[48] In proposing that *sacra doctrina* is a science, Thomas Aquinas is arguing that it is a *practice* that leads to a particular state of mind, as a consequence of which one can habitually reason from cause to effect. In the first question of the *Summa theologiae* he thus states that the unity of the science of theology comes from the fact that it is a single "faculty or habit."[49] This is a helpful reminder that *scientia* is not just a set of ordered propositions, but is also a mental disposition. As Aquinas scholar John

Jenkins has written, the purpose of the science of sacred doctrine is "to induce habits of thought, intellectual habits, in virtue of which a person's knowledge of the cause becomes the cause of, the epistemic grounds for, his knowledge of the effect. Its purpose, that is, is to make one's thinking in a particular field mirror the order of causality."[50] As we can see, this is similar to Albert's notion of mental training. Indeed, for a number of medieval thinkers the "habits" associated with sacred doctrine were not primarily intellectual but practical.[51]

Again we should remind ourselves that the pursuit of philosophical wisdom was not solely concerned with the accumulation of knowledge of the particular subject matter of the various sciences. Equally, it involved becoming a particular kind of person. The attainment of wisdom, as Aquinas noted at the very beginning of the *Summa contra gentiles,* relates to a specific office, where "office" is related to the possession of certain virtues. Aquinas explains, again on the authority of Aristotle, that to discharge one's office is simply to act virtuously.[52] Moreover, for Aquinas, the process of knowing calls for the mind of the knower to become conformed to that which is known. Knowledge of the truth, ultimately identified with contemplation of God himself, thus entails growing into conformity with the divine nature. "The rational creature," as Aquinas puts it, "is made deiform."[53] (Recall also that for Aristotle, contemplation is the activity of God and the contemplative life was possible "in so far as something divine is present in [man]."[54]) As one progresses in knowledge of the first cause, of necessity one acquires "a certain rectitude."[55] What is really going on in the pursuit of both natural philosophy and theology is a process of training and apprenticeship that leads to an intellectual habituation to achieve cognitive restructuring.[56] This understanding of the office of the philosopher as one who undergoes personal transformation will change quite significantly in the seventeenth century, when the philosophical task will be directed outward toward the transformation of nature.

The approach taken by Thomas Aquinas offers us yet another model of how the objects of natural philosophy might have theological import. In becoming familiar with causal relations the knower gains a generic capacity that is transferable to other, more elevated, objects. This conception of "science," in common with the monastic practices of *lectio divina* (an exercise of scripture reading, meditation, prayer, and contemplation) and the symbolic readings of nature of Hugh and Bonaventure, is consistent with the general idea of philosophy as a form of spiritual exercises. "Science" is a handmaiden, not so much because it offers premises for a propositional theology, but because it entails the performance of mental exercises that promote the personal transformation that is the goal of theology. Ultimate truths about divine things can be known only to the properly prepared mind, and the study of nature was seen as contributing to that preparative process.[57]

Medieval Natural Theology?

Natural theology is usually thought of as providing the chief point of contact between the natural sciences and theology throughout history. It is worth asking at this juncture, then, how the approaches to the natural world that we have been considering relate to the project of natural theology. We began this chapter with a reference to Thomas Nagel's proposal that some contemporary analytic philosophers routinely misconstrue the traditional philosophical quest by understanding it primarily in light of their present preoccupations. There is a related kind of misconstrual of the medieval understanding of the theological significance of the natural world. It is tempting for twenty-first-century readers to regard any theologically motivated interest in the study of nature as falling under the rubric of natural theology, understood as an activity that from "neutral" premises offers arguments that establish the existence of God and certain of his attributes.

The analytic philosopher of religion, William Alston, offers this widely accepted definition of natural theology: "the enterprise of providing support for religious beliefs by starting from premises that neither are nor presuppose religious belief."[58] The alert reader will no doubt recall the problematic nature of each of the terms in the conjunction "religious belief" and the potential dangers in applying these terms anachronistically. It is also worth noting at the outset that the expression "natural theology" is almost completely absent from patristic and medieval literature.[59] When we look at the earliest English books we find a similar situation, with a steady increase in usage beginning only from the middle of the seventeenth century. The paucity of references to "natural theology" in the Middle Ages is partly because the parent term "theology" was itself not much in use. On those occasions when the expression "natural theology" does appear, in Augustine and Aquinas, for example, it is used in a negative way. Aquinas thus follows Varro and Augustine in distinguishing three kinds of theology: "natural theology" (*physicum theologiae*), "mythical theology" (essentially euhemerism, the worship of dead heroes), and "civil theology" (state-sponsored worship of images), all of which he regarded as forms of "superstitious idolatry."[60] This association of "natural theology" with pagan thought persisted until the seventeenth century, where we find Francis Bacon alleging that the Platonists corrupted natural philosophy by mixing it with natural theology.[61]

Of course, the relevant activity might be conducted in the absence of the modern label, but for reasons that I hope are now obvious, we ought to resist the temptation to apply "natural theology" to the patristic and medieval approaches to the natural world that I have been describing. For a start, engagement with natural philosophy in the premodern period is thought to offer a particular kind of mental training that will facilitate both moral behavior and theological insight. The process is really one of cognitive formation—a preparation for a form of knowing that remains inaccessible to those who have not undergone the requisite training. This mode of knowing is more like direct experience—tasting and seeing—than logical deduction or inference from premises. Crucially, what one "knows" is not restricted to the fairly limited range of truths associated with natural theology—God, certain of his attributes, immortality, and perhaps postmortem rewards and punishments. Rather, nature offers a parallel revelation that bears testimony to the triune nature of God, to Christ's redemptive death and his resurrection, while at the same time reinforcing a set of quite specific moral prescriptions. To offer a few examples, Hugh of Saint Victor insists that the Trinitarian nature of God can be known through meditation upon the magnitude, beauty, and utility of the creation.[62] Bonaventure was also to contend that the triune nature of God is in some sense inscribed upon his creatures: "The creature of the world is like a book in which the creative Trinity is reflected, represented, and written."[63] Subsequently, the fifteenth-century Spanish theologian Raymond of Sabunde (fl. 1434) was to pen his *Theologia naturalis seu liber creaturarum* (Natural Theology; or, The Book of the Creatures, 1434–36), in which it is argued that the book of nature, like God's other book, was sufficient for salvation and communicated something of God's triune nature.[64] This, clearly, was not natural theology in our sense of the word, and references to the "book of nature" at this time do not refer to an exercise in which premises are sought for arguments from design.

That said, a far more promising candidate for our contemporary version of natural theology can be found in Thomas Aquinas. But even here matters are not straightforward.[65] Aquinas's reputation as the medieval natural theologian par excellence rests in part on the disproportionate attention directed toward the "five ways"—the five so-called arguments for the existence of God that appear in the third article of the second question of the *Summa theologiae*. This brief article is a much-exploited resource for philosophers of religion although it represents a miniscule proportion of Thomas's output. While opinions on the status of the "five ways" vary, a number of commentators have argued that the aim is not to provide a foundation for the whole theological enterprise by establishing, at the outset, the existence of God through reason.[66] Rather, given that we are already assured of the existence of God through revelation, it becomes a matter of explicating the meaning of divine existence in relation to the finitude of the creatures. Philosopher Leo Elders has even offered the intriguing suggestion that the five ways set out a meditative process, and represent stages of the ascent of the mind into deeper levels of understanding the being of God.[67]

When we consider what Aquinas says he is doing in *Summa contra gentiles*—typically regarded as the main work in which he develops a natural theology—again we discover that he is not really engaged in an intellectual activity that begins with premises that require no religious beliefs. Rather he is concerned with premises acceptable to contemporaries who are not committed to the specific truths of the Christian revelation. This is something quite different. The common "rational" grounds from which Aquinas begins, then, are already theistic. Christianity, Islam, and Greek philosophy are understood as already aiming for the same thing—as Aquinas puts it: "almost all philosophy is directed towards the knowledge of God."[68] The apologetic purpose of the argument of *Summa contra gentiles* is to convince readers that Christianity succeeds in achieving these (religious) aims in a way that the others do not. This is why Thomas begins the work with an exploration of "the office of the philosopher" and pursuit of wisdom, for he intends to show that the philosophical quest, as understood by ancient Greeks and Arabs, is one that has a realistic prospect of success only within Christianity. It follows that when Aquinas speaks of "divine truths that reason's enquiry can attain to," this is not equivalent to Alston's "premises that neither are nor presuppose religious belief."[69] Aquinas's starting point presupposes a version of theism, belief in some kind of soul, a teleological conception of human beings that sees certain "ends" as given, and a conception of philosophy understood as the quest for wisdom and spiritual perfection. In short, Aquinas's distinction between modes of truth that surpass the capacity of reason and those that do not maps onto neither the medieval dichotomy of the book of scripture and the book of nature nor the modern distinction between the subject matter of revealed theology and natural theology.

As it turns out, the project of constructing arguments for God's existence based on putatively neutral premises gets under way in earnest in the seventeenth and eighteenth centuries, when a secular conception of reason begins to emerge. It appears in tandem with a new understanding of religion in which propositional beliefs come to play a significant role, and with a rethinking of the goals and purposes of natural philosophy. In short, to relate this development to one of our leitmotifs, we can say that the birth of natural theology accompanies the objectification of the virtues of *religio* and *scientia,* which cease to be qualities of the individual and come to refer solely to activities and bodies of knowledge.

The Discarded Image

One reason that the enterprise of natural theology gains momentum in the early modern period is that the theological modes of understanding nature represented by the symbolic and causal orders described above begin to disintegrate. Their departure makes room for the development of a new "experimental" natural philosophy—one that involves a different, and in some senses more intimate, connection with theology. The gradual demise of symbolic understandings of nature can be accounted for in a number of ways, but one factor was growing skepticism about the allegorical mode of reading scripture and nature. For its patristic and medieval advocates, allegorical interpretation, as we have already noted, was not simply a matter of reading multiple meanings into words, but, as we see in Augustine, assumed a view of the world in which natural objects had spiritual meanings. As a consequence, any attack on the allegorical interpretation of scripture would necessarily have implications for how the world was interpreted, since the significance of words and of things was intimately connected in a rich web of symbolic meanings. Reservations about the abuses of allegorical reading begin as early as Nicholas of Lyra (1270–1349) who, while not dispensing with the spiritual senses of scripture, insisted that the literal sense was foundational. But it was the Renaissance and Reformation that spelled the beginning of the end for allegory as a general interpretative stance that could be applied to scripture and nature.[70]

Martin Luther bluntly declared that allegory was for "weak minds" and "idle men." For him it was the literal sense that contained "the whole substance, nature and foundation of the holy scripture."[71] John Calvin also prioritized the "historical" or "literal" sense of the text, and chastised those who pursued allegory.[72] This censure of allegory was motivated by the Reformers' vesting of religious authority in scripture alone. The principle of sola scriptura, in turn, presupposed that when it counted, the meaning of scripture was clear and unambiguous. The Reformers also sought to liberate the text of scripture both from the interpretative authority of the Catholic magisterium and from a long exegetical history that in their view had obscured the true meanings of the text. The critique of allegory was accompanied by a general suspicion of symbolic representations, and thus the program of the Reformers was associated, to varying degrees, with a contraction of the sacramental world, a sometimes virulent iconoclasm, and a corresponding positive emphasis on the unvarnished written word.

Natural philosophers in the seventeenth century gave clear expression to the new nonsymbolic understanding of nature that went hand in hand with these new readings of the book of scripture. Commenting on the medieval idea of the book of nature in his 1605 manifesto for a new kind of natural science, Francis Bacon observed that "for as all works do shew forth the power and skill of the workman, and not his image; so it is with the works of God which do shew the omnipotency and wisdom of the maker, but not his image."[73] The idea that the world shows the image of God, Bacon goes on to say, is a "heathen opinion." To some degree this was true, for it was Plato, after all, who had asserted that the world is "a sensible God who is the image of the intellectual."[74] But as we have seen, this view also informed a Christian understanding of the book of nature. Bacon's stance contrasts directly with that of Origen, for whom "all created things have a resemblance and likeness to heavenly things," and Bonaventure, for whom "the creature of the world is like a book in which the creative Trinity is reflected, represented, and written."[75] It was also in opposition to contemporary expressions of the emblematic worldview inspired by Platonic hermeticism.[76] Bacon's denial of the symbolic transparency of the natural world is accompanied by his recommendation of a new method for revealing "the true signatures and marks set upon the works of creation" as opposed to what he calls the "idols of the mind" and "empty dogmas" of traditional natural philosophy.[77]

It is important to understand that Bacon, like many advocates of the "new philosophy," did not imagine himself to be stripping the universe of its religious significance. Rather he presents himself as offering a genuinely Christian approach to nature, in comparison to preceding approaches that were understood to have been contaminated by pagan philosophy (hence Bacon's characterization of them as "heathen"). This was a reformation of the sciences that was to parallel the reformation of religion that, from a Protestant perspective, was similarly about the purification of paganized Catholicism. Bacon proposes a new, nonallegorical way of reading the book of nature. This was characteristic of other pioneers of the new sciences. Kepler and Galileo, for example, retained the notion of the book of nature, but contended that the language in which it was written was mathematical.[78] Rather than a vast canvas of theological symbols, the cosmos instead bore inscriptions written in the spare and precise language of geometry and mathematics. As Galileo famously expressed it:

Philosophy is written in this grand book, the universe, which stands continually open to our gaze. But the book cannot be understood unless one
first learns to comprehend the language and read the letters in which it is
composed. It is written in the language of mathematics, and its characters
are triangles, circles, and other geometrical figures without which it is
humanly impossible to understand a single word of it.[79]

It was not a contemplative, hermeneutical practice that would lay bare
the hidden intelligibility of the cosmos, but rather mathematical skill.
Nature, on this model, consists of idealized geometrical entities rather
than divinely instituted symbols. While such a view is often thought to be
a Platonic one, a key element of the Platonic conception of the cosmos —
its intrinsic moral significance — was missing. Plato's assumption of the
edifying qualities of mathematical reasoning is also increasingly absent.

The specific form that this mathematical order takes will be a new
one — the idea that there are natural laws, capable of mathematical formulation, and imposed on matter directly by God. René Descartes (1596–
1650) put it this way: "God imparted various motions to the parts of matter when he first created them, and he now preserves all this matter in the
same way, and by the same process by which he originally created it."[80]
Motion, then, is not a feature of matter itself, but depends upon the divine
will. This conception of nature as obeying arbitrary laws conferred by God
would help to motivate empirical inquiry into the operations of nature. As
Descartes explained: "Since there are countless different configurations
which God might have instituted here, experience alone must teach us
which configurations he actually selected in preference to the rest."[81]

If we turn from the physical world to living things, we see another new
version of the book of nature metaphor. Robert Boyle (1627–91), one of the
leading lights of the newly established Royal Society, believed that reading the book of nature called for procedures that could penetrate beneath
the visible surface of things, not in order to disclose their transcendental referents, but to reveal their remarkable physical design. Boyle spoke
of the creatures as "texts, to whose exposition physiology is necessary."
Their meanings, if we may still call them that, lay within them, rather
than beyond them. The discernment of their meanings was to be facili-

tated through experiment, dissection, and artificial magnification. In his characteristically labored prose, Boyle explained that living things were "texts" whose interpretation called for "penetrating indagations" directed toward the discovery of their "unobvious properties." Alluding to the practices of shorthand, then much in vogue, Boyle also referred to the objects of nature as "the stenography of God's omniscient hand."[82]

Even more direct evidence of the demise of the symbolic world on which the older metaphor had depended comes from works of natural history that identified themselves with the reformed approach to nature characteristic of the new experimental science. Whereas traditional natural histories, such as *The Historie of Foure-Footed Beastes* (1607) of the clergyman Edward Topsell (ca. 1572–1625), still enumerated the symbolic meanings of the creatures, this emblematic stance was eschewed by advocates of the "new philosophy."[83] Nehemiah Grew (1641–1712), a pioneer of plant physiology and active early member of the Royal Society, made a point of excluding "Mystick, Mythologick, or Hieroglyphick matter" from his catalog of the Royal Society's natural history collection. Instead, he would be focusing on "the Uses and Reasons of Things."[84] In his *Ornithology* (1678), coauthored with Francis Willoughby, John Ray similarly announced that he would not be dealing with "*Hieroglyphics, Emblems, Morals, Fables, Presages* or ought else appertaining to *Divinity, Ethics, Grammar,* or any sort of Humane Learning." These were not deemed to have a proper place in a legitimate natural history.[85]

These developments signal the death of a universal hermeneutical framework in which the books of scripture and nature were interpreted together. Now, even the book of nature was subject to a plurality of hermeneutical practices—mathematical, anatomical, taxonomic. But the collapse of the unified system of interpretation, and the separation of the study of texts and the natural world, by no means implied that the study of nature was to be pursued independently of theological considerations. Instead a new partnership developed between theology and the new science.

Lost Causes

Accompanying the evacuation of meaning from natural objects was the stripping away of their intrinsic causal powers.[86] The predominant scholastic position on causal order was a "bottom up" understanding that accounted for change in the world in terms of the inherent powers and virtues of its constituents. This view derived ultimately from Aristotle, for whom the powers of things derive from their forms.[87] To this basic Aristotelian conception the scholastics had added the idea of a creator God, who is involved because he is responsible for the being of those objects. Thus every event has a primary cause (God) and a secondary cause (the powers of the individual agent).[88] Some of the later scholastics had already expressed reservations about this view of things, but it was the advent of Descartes's new notion of laws of nature that spelled the beginning of the end for intrinsic powers and qualities.[89] For Descartes, as we have already seen, it is laws of nature and not intrinsic properties that offer the best prospect for explaining the operations of the natural world. These laws of nature were not descriptive of relations among the properties of objects, but rather of divine volitions.[90] On this "top down" understanding, the regularities of nature were directly imposed upon it by God. Such a view not only discounted the causal role of Aristotelian forms, but ultimately led to doubts about whether there were any genuine secondary causes in nature at all. This tendency toward occasionalism—the idea that God is the only true cause of natural events—culminated in David Hume's (1711–76) suggestion that causation was essentially psychological rather than ontological, in human minds rather than the world.[91] Occasionalism also fitted well with the idea of God promoted by the Protestant Reformers, who emphasized the primacy of the divine will and stressed God's omnipotence and transcendence.

Somewhat paradoxically, then, the pious idea that God was the only genuine cause in the cosmos and that natural objects had no causal powers of their own led to the direct equation of divine and natural causality. For a number of key seventeenth-century natural philosophers, and in particular the Newtonians, the regularities of nature were a manifestation of the continuous and direct activity of God.[92] The Newtonian philosopher-

theologian Samuel Clarke thus maintained that there was no such thing as
"the power of nature" or "the course of nature." There was, rather, "nothing else but the will of God producing certain effects in a continued, regular, constant, and uniform manner."[93] In this collapse of the distinction
between natural and supernatural causation lay the seeds of a thorough-
going naturalism, for once divine activity was placed on the same level as
natural activity the operations of nature could be understood as having
either divine causes, or natural causes, but not both at once. Causal explanation became a zero-sum game, and the disjunction God-or-nature
would increasingly be resolved in favor of the latter. Some commentators
have suggested that this tendency to place divine causation on a par with
natural causation is to be attributed to the influence of John Duns Scotus.[94]
But whatever its source, it is certainly characteristic of understandings
of causation among many early modern natural philosophers. In much
the same way that symbolic meanings of nature and scripture were col-
lapsed into a single literal sense, the various causal layers of Aristotelian
scholasticism came to be flattened into a single layer of univocal efficient
causes. There was one order of meaning—the literal sense—and one level
of causation—efficient causation. Both of these developments were promoted by religious thinkers, and for religious reasons. They were both a
precondition for the emergence of modern science. Yet, the ultimate effect
of this flattening of the scope of meaning and causation was that modern
science and theology would come to occupy the same explanatory territory, and this established the conditions for competition between them.

The fading fortunes of the Aristotelian understanding of causal order
extended not only to the powers of inanimate objects, but to human
powers and virtues as well. As we saw in the first chapter, virtues were
understood as habits that perfect natural powers. Natural powers, in turn,
move the entities in which they inhere—in this case human persons—
toward their natural ends. This was how Aristotelian "final causes" were
thought to operate. The Protestant Reformers objected to key elements of
this conception of human habits and ends. For a start, they took a much
harder line than their scholastic predecessors on the moral and cognitive
incapacity that resulted from original sin. One consequence of this un-
compromising stance was a skepticism about whether unaided human

reason could penetrate to the essences of things, or know about their true ends, or discern the intentions that God might have had in creating natural things. These doubts, in turn, were a motivating factor in a new experimental natural philosophy that was more critical in its methods and more modest in its ambitions than the Aristotelian natural philosophy that it was destined to replace. The Reformers, and Luther in particular, were also sharply critical of the Aristotelian notion of *habitus*, because of its perceived complicity in a mistaken view of justification, the doctrine that explained how sinful human beings might be reconciled with God. Not only were flawed human beings incapable of discerning their true end, but even if they were, their weakened moral and intellectual powers would be totally inadequate to move them in the right direction. These critical assaults, considered in more detail in the next chapter, would necessitate a rethinking of the foundations of Aristotelian moral philosophy, and of the role and operations of the moral and intellectual virtues. As a consequence, *scientia* and *religio* will begin to assume the shape of the modern categories "science" and "religion."

[4]

Science and the Origins of "Religion"

An "inner process" stands in need of outward criteria.
—Wittgenstein, *Philosophical Investigations*, §580

The Christian religion has very strong evidences.
—Samuel Johnson, in Boswell, *Life of Johnson* (1787)[1]

The Chief End of Science is to beget Virtue.
—John Sergeant, *Method to Science* (1696)

Recent years have witnessed the publication of a significant number of works purporting to offer scientific explanations of religion, often in terms of cognitive science or evolutionary psychology.[2] In an often undistinguished field, David Sloan Wilson's *Darwin's Cathedral* (2002) represents a relatively sympathetic evolutionary account of religion. One of the besetting difficulties of this genre lies in the slippery nature of the putative object of study—religion. Wilson addresses this problem by suggesting that catechisms, understood as concise and systematic presentations of religious beliefs, offer a good way of capturing the essence of religion. Catechisms, he informs us, "are a gold mine of information for the evolutionary study of religion"; they are "'cultural genomes' containing in easily replicated form the information required to develop an adaptive community"; they are "short enough for detailed analysis" and they enable "the comparative study of religious organizations." He concludes that it is hard to imagine a better historical database than the information provided by the content of catechisms.[3] Crucially, for Wilson, catechisms lend themselves to scientific and comparative analysis because they are "measurable aspects of the world."[4]

That such an approach is now plausible is testament to a remarkable change in the understanding of both religion and science that can be traced back to the early modern period. What began to take place then was that the philosophical exercises and bodies of knowledge employed in the inculcation of the interior virtues of *scientia* and *religio* came to stand in for the things themselves in their entirety. The content of catechisms that had once been understood as techniques for instilling an interior piety now came to be thought of as encapsulating the essence of some objective thing—religion. Religion was vested in creeds rather than in the hearts of the faithful. In a related process, the label "*scientia*," which had traditionally referred to both a mental disposition and a formal body of knowledge, came to be associated with the latter alone, eventually giving rise to an objective thing—science. While there had once been a close correspondence between science considered to be a virtue and science understood in terms of demonstrable knowledge, from this period onward, science was increasingly thought of as a body of systematic knowledge or a method that existed quite independent of the dispositions of its practitioners. The very possibility of a "science of religion" rests upon these transitions, and indeed the work of David Sloan Wilson is a distant ancestor of a new approach to religion that begins in the seventeenth century.

The End of Ends

The processes that led to these modified understandings of *scientia* and *religio* are various, but the fact that both undergo similar transitions is suggestive of at least some common causes. One reason for this reconfiguration is that the prevailing Aristotelian model of virtues—understood both as moral qualities of the individual and as inherent propensities of natural bodies—was called into question. In the moral sphere, the Aristotelian scheme had been one in which certain human ends were taken as given, and the virtues were ways of promoting those ends. Reason informs us of our true end and the means to achieve it. The task of ancient philosophy had been to assist the individual in the cultivation of those habits that would help them achieve their true end.[5] With the inception of Christianity the moral cultivation that had formed the central focus of philosophy was moved into the religious domain, and distinctively Christian elements were added to the basic scheme. Among these were the concept of sin, the theological virtues (faith, hope, and charity), and the idea that ethical precepts should be understood both as divine commands and teleological injunctions (that is, prescriptions that promoted the attainment of natural human ends). Yet the basic Aristotelian structure was retained through the Middle Ages. In the sixteenth and seventeenth centuries, however, the foundations of this framework began to crumble. Protestant critics found fault with the relationship between habit and merit, and with assumptions about the capacity of reason to determine human ends and the means to accomplish them. In the sphere of natural philosophy, the Aristotelian understanding of virtues, qualities, and ends as intrinsic properties of natural bodies was also subjected to searching criticism. Change in the natural world came to be understood in terms of laws of nature that were thought to operate externally on inert particles of matter.

One objection to the prevailing understanding of the virtues, mounted by Protestant thinkers, was that it implied that genuine moral credit could be acquired by repeated practice. According to Martin Luther, the Aristotelian position amounted to the view that "he who does much good will thereby become good." In Luther's judgment, this view of moral advancement—a doctrine of salvation by good works—had corrupted medieval Christianity.[6] Aristotle and Aquinas were disparaged as "fools" for thinking, as he put it, "that the virtues of love, chastity, and humility can be achieved through practice."[7] For Luther this stance was at odds with the New Testament teaching that no amount of human effort is sufficient to make one good. Practice does not make perfect. Rather, God, through his grace, treats us *as if* we were perfectly righteous. Luther thus contended that righteousness "is not in us in a formal sense, as Aristotle maintains, but is outside us."[8] Calvin agreed that the teaching of Saint Paul was that righteousness is not "a quality infused into us." The New Testament, he pointed out, makes no reference to good works but "tells us that righteousness must be sought without us."[9]

Luther and Calvin may have overstated the affinities between the scholastic views and those of Aristotle, particularly on the issue of the role of human efforts in the cultivation of the virtues. But what was distinctive about their position in relation to the scholastics was their contention that grace should be understood in relational rather than substantial or metaphysical terms.[10] To put it another way, individuals were not justified by virtue of an inner quality, whether acquired or infused, but on account of their relationship to God. That relationship, in turn, was determined solely by the divine will. As we will see shortly, an analogous change was subsequently to take place in the sphere of natural philosophy, with patterns in the natural world being understood no longer in terms of the inner qualities of things, but rather as ordered relationships imposed on the world directly by the will of God. In the spheres of morality and natural philosophy, laws of nature thus came to assume the role previously played by inherent virtues.

A closely related element of the Reformers' denial of key elements of the Aristotelian program was their skepticism about the possibility of perfecting human nature. On the Aristotelian model, both the moral and intellectual virtues promoted the perfection of particular natural powers.[11] However, the premise that human beings could perfect any of their natural powers was inconsistent with the Protestant understanding of original sin. When, for example, English poet John Milton represented the pagan understanding of the virtues in *Paradise Regained,* he noted that the philosophers had been ignorant of "how Man fell" and that because of this they had mistakenly sought virtue "in themselves."[12] It was not merely moral capacities that had been damaged by the Fall, for the consequences of original sin extended to the intellectual faculties as well, thus encompassing both *religio* (a moral virtue) and *scientia* (an intellectual virtue). Understood in this light, Calvin's notorious doctrine of "*total* depravity" was less about the extreme nature of the depravity suffered by human beings than about the fact that depravity extended to every human faculty, including the intellectual. The effects of original sin were not confined to the moral faculties, as some of the scholastics had taught.[13] In short, Protestant understandings of divine grace and original sin were inconsistent with the idea that certain habits could perfect the moral and intellectual powers.

Yet another way of seeing the impact of these new ideas is to consider how they eroded the Aristotelian matching of natural ends with particular human propensities. For many post-Reformation thinkers—both Protestant and Catholic—the disordering of nature that followed the sin of Adam and Eve had disturbed the original alignment of natural human ends with the means to achieve them. Aristotle had maintained that all men by nature desire to know. This propensity he took to be a natural and distinguishing feature of humankind, and he concluded that the highest activity for human beings lies in the fulfillment of that desire, which is to say, in intellectual contemplation.[14] But for a number of Aristotle's early modern detractors this analysis failed to take into consideration the distorting effects of human sin. For them, human nature as we presently encounter it does not represent the natural and original condition of human beings. Accordingly, the data provided by an examination of our "natural" propensities may not in fact point to our natural end, but may instead simply exemplify our corrupt condition. Our apparently natural inclination to seek knowledge might thus be a culpable curiosity—an originally innocent desire, now disordered by sin.

Aristotle had also thought that if entities had a natural propensity toward some end, then nature (or, on the scholastic reading, God) would have provided the means to achieve it. In the case of human beings, our natural inclination toward a scientific knowledge of things was thus a sure indication that we possessed the cognitive capacity to arrive at that kind of knowledge since, as Aristotle constantly reminds us, nature does nothing in vain. This Aristotelian position informed the theories of knowledge adopted by the scholastics. John Duns Scotus put it this way: "If the proper operation of the intellect . . . is to know the thing which is true, it seems hardly fitting that nature should not endow the intellect with what is prerequisite for such an operation."[15] The application of this principle to the intellectual virtues gave rise to the view that, all other things being equal, the exercise of *scientia* (along with *intellectus* and *sapientia*) will normally give rise to reliable knowledge. For this reason, Aristotelian natural philosophy was based on commonsense generalizations drawn from everyday observations of the natural world.

For those who took seriously the deleterious effects of the Fall, however, there were no guarantees that our corrupted faculties, operating in a postlapsarian world, would naturally point us in the direction of truth. On the contrary, they were more than likely to lead us into error. It was conceded that there might once have been a natural correspondence between the mind and the world, but this capacity of the mind to accurately represent the world was now regarded as one of the chief casualties of the primeval Fall. Francis Bacon, echoing the Pauline sentiment that in the present world we see "through a glass, darkly," described the mind as a distorting mirror that routinely deceives us. Bacon's ambitious project to set science on new foundations was motivated by this basic question: "whether that great commerce between the mind of man and the nature of things . . . might by any means be restored to its perfect and original condition."[16] In seventeenth-century England, exponents of the new experimental approach to philosophy stressed the fallibility of human reason, the impossibility of penetrating to the true essences of things, the need for repeated experimental observations carried out under special conditions, and the necessity for scientific researches to be concerted, collective, and cumulative. Exemplifying this general attitude, Robert Hooke, first curator of experiments at the Royal Society, wrote of the human condition that "every man, both from a deriv'd corruption, innate and born with him, and from his breeding and converse with men, is very subject to slip into all sorts of errors." He concluded that "these being the dangers in the process of humane Reason, the remedies of them all can only proceed from the real, the mechanical, the experimental Philosophy."[17]

So it was that an "unnatural" and somewhat counterintuitive regimen of experiments was designed to interrogate a fallen nature with the appropriate level of skepticism, while at the same time addressing the human predilection for delusion and self-deception.[18] As for the content of natural philosophy, it would be premised on a view of the world similarly couched in nonteleological terms. As noted earlier, replacing the teleological causal powers that moved objects toward their natural ends were divinely imposed laws of nature that paralleled, in certain respects, the divine imperatives that constituted the moral law. Again, then, divinely authored laws of nature would replace the inner virtues or qualities of natural things, while divinely authored moral laws would similarly replace human virtues.[19]

Related to this skepticism about the natural knowledge-making propensities of human beings were various levels of denial about the possibility of determining the true ends of natural things, and even, to some extent, the ends of human life. Thomas Hobbes (1588–1679) declared in the *Leviathan* (1651) that "there is no such *finis ultimus* (utmost aim), nor *summum bonum* (greatest good), as is spoken of in the books of the old moral philosophers." English poet and satirist Samuel Butler (1612–80) agreed that "it is very probable we do not understand (although we may believe) the Purpose for which wee were created, for if we are ignorant of the immediate, and nearest Causes of our selves, much more must wee be so of the most remote."[20] Our supposed ignorance of the true ends of other aspects of nature had already been championed by Francis Bacon and René Descartes. Bacon thought that "the final cause rather corrupts than advances the sciences." Descartes agreed that because human nature is "very weak and limited whereas the nature of God is immense," it follows that "the customary search for final causes [is] totally useless in physics."[21]

Admittedly, strong claims continued to be made during this period about the true ends of human life. As well-drilled generations of Presbyterian children could once have testified, the first article of the Westminster Catechism (1647) declares that "Man's chief and highest end is to glorify God and enjoy him forever."[22] But these confident creedal assertions about the ends of human life were not grounded in an analysis of human desires or powers as they had been for Aristotle, but were justified by direct appeals to scripture. (This is the import of Butler's claim that we may have a belief about the purpose for which we are made, even though we do not understand it.) The true ends of human life thus needed to be given through revelation, and the means to achieve them were set out in the divine commands laid out in scripture.[23] In fact, just as the critique of Thomist-Aristotelian teleology is informed by an Augustinian view of original sin and its consequences, so it is Augustine's view of human ends,

along with his distinction between use and enjoyment that is now re-
instated: the ends of human life are to enjoy God and to use the creatures.
In his commentary on the Westminster Catechism, Presbyterian minister John Flavell thus reiterated that the chief and highest end of man was
to glorify God and enjoy him forever, adding that his secondary end was
"prudently, soberly, and mercifully, to govern, use, and dispose of other
Creatures in the Earth, Sea, and Air, over which God gave Man Dominion."[24] As we shall see in more detail in the next chapter, the motifs of self-
cultivation and self-mastery that had been central to classical accounts of
philosophy, and which to some degree had found their way into medieval
understandings of human ends, would now be directed outward onto the
world. The goals of human life will be placed within a pattern of redemption history in which the way back to God involves reestablishing mastery
of the natural world that had been lost as a consequence of the Fall. This
realigned understanding of true human ends thus eroded the significance
of the virtues of *religio* and *scientia* by undermining the assumptions upon
which they rested. At the same time, it led to the establishment of a set of
religious justifications for the practical dominion over nature that would
motivate the new sciences of the seventeenth century and set the scene
for the emergence of the modern idea of science as a set of techniques directed toward mastery of the natural world.

None of this means that natural philosophy has now been fully extri-
cated from the moral realm. It is rather the beginning of a decoupling
of knowledge-making enterprises from a particular understanding of the
virtue of *scientia*. Insofar as knowledge continues to be related to the moral
subject, it might be annexed to such virtues as charity (Bacon) or gener-
osity (Descartes). Neostoicism also promoted its version of the virtues,
and these, too, could be related to knowledge-making enterprises in various ways, and without the foundations of an Aristotelian teleology.[25] More
generally, virtues were often redescribed in terms of whether they yield
behaviors that accord with positive laws. John Locke, like Machiavelli and
Descartes before him, dismissed the classical virtues and vices as mere
"names pretended and supposed everywhere to stand for actions in their
own nature right and wrong." In the absence of any consensus about naturally evident human ends, the virtues seemed simply to be arbitrary in-

ventions of the ancients. Any legitimate conception of virtue and vice, for Locke, had to be reformulated in terms of moral law, the primary source of which is "the will of a supreme invisible law-giver." In Locke's own words: "By whatever standard soever we frame in our minds the ideas of virtues or vices . . . their rectitude or obliquity consists in the agreement with the patterns prescribed by some law."[26] On his model, "virtues" are now understood in terms of the behaviors they generate—as one commentator has recently put it, they are explained "from the outside in."[27] In the case of religion, the virtue of *religio* will increasingly be understood from the outside in, as the modern ideas of religion and the religions make their first appearance.

This revised understanding of the moral virtues directly parallels developments in the new experimental approach to the physical world, which concerns itself with the behaviors of objects in terms of their conformity to laws, rather than in terms of supposed internal virtues and tendencies. Indeed, the religious critique of the idea of *habitus* and its relation to the virtues was reinforced by elements of the new natural philosophy. This is hardly surprising, since in the Aristotelian scheme of things both physical and moral action operated along similar general principles. Descartes, for example, opposed the idea of *habitus* because it had implications that were inconsistent with the mechanical philosophy that he espoused.[28] *Habitus* was intimately linked to the scholastic idea of forms and qualities that he had sought to expel from the universe. By the same token, those with a stake in maintaining the traditional justifications of the moral virtues were alarmed that their foundations might be undermined by the demise of Aristotelian natural philosophy. The vehemence with which some of Descartes's theological opponents attacked his natural philosophy provides us with a measure of what was at stake. His more able detractors saw clearly that doing away with the natural powers of things would provide significant challenges to the prevailing understanding of the moral powers of human agents. Descartes's most vocal theological critic, the Dutch Calvinist Gisbert Voetius (1588–1676), pointed out that the denial of the existence of the inherent powers would render moral explanations in terms of an inner *habitus* completely meaningless. A coherent account of moral action, in his view, had to be grounded in Aristotelian philosophy. "I should like to see," he declared, "how those who deny the existence of natural faculties or potencies could defend [the concept of] *habit*—the necessity of which is urged by Scripture and reason alike."[29] In short, the fate of the inherent virtues of natural things was intimately related to the fate of the moral and intellectual virtues. For this reason, as a consequence of challenges to Aristotelian understandings of virtues both natural and human, *scientia* and *religio* took on new meanings, and were increasingly associated with systems of thought and belief in the familiar modern sense.

Definite Articles

Another reason that *religio* came to be understood "from the outside in," as it were, was that the religious reformers of the sixteenth century were insistent that religious faith be "explicit." A prominent feature of the rhetoric of Protestant Reformers was their denigration of traditional notions of implicit faith—the idea that explicit knowledge of the more abstruse doctrines of Christianity need not be enjoined upon the laity.[30] The Reformers insisted instead that Christian believers be able to articulate the doctrines they professed, and do so in propositional terms. In his *Institutes,* John Calvin thus decried ignorance of the doctrines of Christianity:

> [The] true religion which is delivered in the Scriptures, and which all ought to hold, they readily permit both themselves and others to be ignorant of, to neglect and despise; and they deem it of little moment what each man believes concerning God and Christ, or disbelieves, provided he submits to the judgment of the Church with implicit faith.[31]

Explicit belief was certainly not intended to be a substitute for personal piety—what Calvin refers to as "training in godliness." Rather the idea was that such training necessarily involves some familiarity with doctrines. While Calvin links such knowledge with what he calls "true religion" (*veram religionem*), tellingly, this was typically translated "*the* true religion." This rendition appears as early as the first English translation of the *Institutes* (1561).[32] This subtle insertion of the definite article signi-

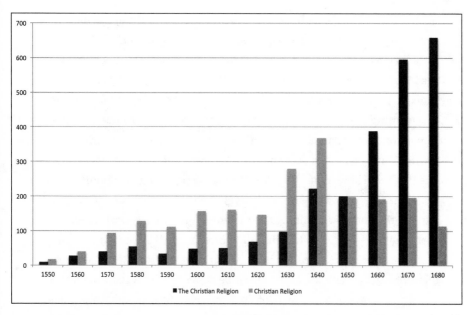

Figure 3. Relative frequency of the expressions "Christian religion" and "the Christian Religion" in English books, by decade, 1550–1680.

fies an important change in how religion is conceptualized. For Calvin, the profession of explicit beliefs is directed toward the promotion of an inner quality — "true religion." But the expression "*the* true religion" places the primary focus on the beliefs themselves, and religion thus becomes primarily an existing thing in the world, rather than an interior disposition. Tangible evidence of this profound shift in the meaning of "religion" can be seen in the increasing frequency of the deployment of the definite article before "religion." In English books printed during the first decade of the seventeenth century, the expression "Christian religion" (without the definite article) is used five times more frequently than "*the* Christian religion." By the final decade of the century the latter expression is much more common (see figure 3). A largely unintended consequence of an insistence on explicit belief and creedal knowledge was thus the invention of *the* Christian religion, constituted by beliefs. Henceforth both Protestant and Catholic reform movements will emphasize the importance of doctrinal knowledge, with the consequence that propositional beliefs become one of the central characteristics of the new "religion."

Linked to this new emphasis on explicit belief was a movement toward the instruction of the laity through a plethora of printed materials: collections of sermons, devotional writings, Psalters, and exegetical writings. The printing press thus made its contribution to the new understanding of religion. Protestant clergy in particular stressed the importance of the inculcation of religious doctrines through catechesis—a venerable practice that they claimed had been lost to the church during the Middle Ages. In the preface to his Large Catechism (1530) Martin Luther enjoined upon parents the duty of examining their children on a weekly basis to ensure that they were diligently learning their catechisms.[33] So popular was this method of Christian education in England that by the middle of the seventeenth century there are reckoned to be some five hundred catechisms in print.[34] Again, these digests of doctrines could be thought of both as promoting the personal quality "Christian religion" *and* as being constitutive of something called "the Christian religion." One of England's leading sixteenth-century Puritan divines, William Perkins (1558–1602), maintained that the Apostles' Creed embodied "the very pith and substance of Christian religion, taught by the Apostles, imbraced by the ancient fathers, sealed by the blood of martyrs."[35] Consistent with the claims made by David Sloan Wilson at the beginning of this chapter, catechisms came to be thought of as embodying the content of "the Christian religion." Indeed, historian Jean Delumeau has gone so far as to contend that with the attempt to extend to the general populace a religious discipline that once had been the preserve of monastic communities, we see the first full-scale "Christianization" of Europe. According to Delumeau, this process—a consequence of the reformations of both Luther and of Rome—peaked in 1700.[36] This thesis of a delayed Christianization of Europe is only plausible, however, if we have in mind the identification of Christianity as a religion, where the latter is equated with neither an inner piety nor participation in the rites of the church, but is rather understood as the outward profession of a set of propositional beliefs.

Turning to understandings of "science" during this period, it is possible to discern parallel shifts in emphasis. As already noted, the idea of science as a mental habit persisted into the seventeenth century. A standard Latin scholastic textbook of the period—one that Descartes, for example, drew upon—states that *scientia* is usually defined as a "condition of the mind" (*habitus mentis*) that is reliable, stable, and clear. The author observes a clear distinction between this mental disposition and "actual knowledge" (*actuali scientia*). Latin dictionaries tell a similar story. In his *Lexicon Philosophicum* (1613), Rudolph Goclenius gives two senses of "*scientia*"—first, "a habit acquired through demonstration," and, second, "knowledge acquired through demonstration." In what is perhaps an indication of the tenor of his time, he specifies later in the entry that the proper use of the term relates to the knowledge, rather than the habit.[37] We find a similar treatment in English textbooks. Robert Recorde, the Welsh mathematician and physician (notable for having been the first to introduce the equals sign into algebraic formulations), gave this definition: "Science is an habit, (that is) a ready, prompt and bent disposition to doe any thing, confirmed and gotten by long study, exercise and use." Later in the century fellow Oxonian Robert South would speak similarly of "an Act of *Demonstration* producing a *Habit of science* in the Intellect."[38] Common to these examples is the idea that the purpose of study is to instill intellectual virtues. The assumption behind these traditional understandings of *scientia* is that the study of logical reasoning and the rehearsal of mathematical demonstrations will produce *scientia* in the individual in much the same way that catechesis will instill *religio*. Focusing as they did on logical demonstration and devices such as the syllogism, these scholastic methods were susceptible to the interpretation that they were incapable of generating new knowledge. However, learning, traditionally conceived, had not aimed at new discovery, but was rather a movement toward a fulfillment of a natural desire to know. This, in turn, was understood in light of the assumption that the final end of human existence (and divine existence for that matter) lay in contemplation.

The sixteenth and seventeenth centuries, it must be said, certainly do not witness the complete disappearance of the interior dimensions of science and religion. But these habits of mind are now often pursued as a means of accomplishing exterior ends. The virtues of religion and science are upheld insofar as they are considered to have a social impact and pro-

mote the improvement of human society. Thus, in addition to its contributions to the reification of religion, catechesis was intended to promote
a collective discipline aimed at the reformation and transformation of society. It was the external expression of religion evidenced in morally disciplined behavior that would contribute to an ordered community. This
logic was particularly prevalent in Calvinist territories. The philosopher
and historian Charles Taylor has spoken in this context of "the rise of a
disciplinary society," and this feature of early modern European societies
has been plausibly linked by a number of commentators to the appearance
of the modern state.[39] Historian Michel Foucault has also pointed out how
the disciplinary society was manifested "externally," in material and spatial terms, speaking of "the regime of disciplinary power" and the manner in which discipline proceeded "from the distribution of individuals
in space." This spatial element of discipline was evidenced in the confine-
ment of vagabonds and paupers, the imposition of the monastic model
on schools, the construction of military barracks and prisons, and, in due
course, the appearance of the industrial workshop.[40]

   The rise of the disciplinary society was also linked to the emergence of
new scientific communities—groups of individuals whose activities were
governed by experimental protocols, and whose goal was the betterment
of human society. Communal and disciplined scientific activities, increasingly confined to a specific site, the laboratory, thus became the external
manifestation of the inner virtue of *scientia*. The justification for pursuing
"science" in this new sense would move from the personal moral sphere
to a broader social context, albeit one that retained a central moral and
religious focus. Francis Bacon exemplifies this development in his contention that knowledge should not be divorced from action and production
"as if there were to be sought in knowledge a couch, whereupon to rest a
searching and restless spirit; or terrace, for a wandering and variable mind
to walk up and down with a fair prospect . . . and not a rich store house, for
the glory of the Creator and the relief of man's estate."[41] Scientific knowledge, he says elsewhere, should be sought not for "the quiet of resolution"
but for "a restitution and reinvesting (in great part) of man to the sovereignty and power . . . which he had in his first state of creation."[42] This is
a new understanding of what *scientia* entails. In the course of time, as was

the case with *religio,* the external manifestations came to take the place of the whole, almost totally eclipsing the interior elements that had once been directed toward the perfection of the human intellect.

Religion and Religions

One important difference between the trajectories of *religio* and *scientia* lies in the fact that while in the Middle Ages there were "sciences," understood as discrete bodies of systematic knowledge, there were no plural "religions" (except in the narrow sense of different monastic orders). A striking feature of the changing grammar of "religion" in the early modern period is that along with the introduction of the article "the" (*the* Christian religion), the plural term "religions" also comes into wide use. Again, creeds and catechisms were to play a role in this development. While they may initially have been promoted as devices for fostering religious formation, confessional statements could be put to other uses, too, most significantly in the drawing of territorial boundaries. The Peace of Augsburg (1555), for example, sought to resolve conflicts between the Holy Roman emperor Charles V and the Schmalkaldic League, an alliance of Lutheran princes. The settlement was momentous, because it provided for the permanent division of the Holy Roman Empire, premised on the idea that religious differences could be given objective formulation and that the inhabitants of particular territories could be identified on the basis of their religion. The conditions of the treaty specified two "religions"—"the old religion and the Augsburg Confession" (i.e., Catholicism and Lutheranism). Lutheranism thereby gained the status of a religion, and was defined in terms of the twenty-eight articles of the confession that represented Lutheran belief and practice. The principle of these territorial divisions came to be encapsulated in the Latin phrase *cuius regio, eius religio*— "whose land, his religion"—which upheld the idea that the religion of the prince would become the official religion of the region. For the first time, "religion" could be understood as a political and legal construct. As a measure of this, various religious "tests" were developed that would establish whether or not one genuinely subscribed to one religion or another.[43]

Other "religions" were to follow. The Augsburg Peace was inherently

unstable because of its proposal that "all such as do not belong to the
two above named religions [Catholic and Lutheran] shall not be included
in the present peace but be totally excluded from it."[44] This meant that
Calvinists and others in the reformed confessions did not belong to the
two specified religions and were not afforded protection under the treaty.
On one common view, the inadequacies of the Augsburg settlement led to
the disastrous "wars of religion" that ravaged Europe for decades. In this
version of events, religiously fueled violence was only brought to an end
by the formation of the modern state, which banished religion to the private sphere, limiting its capacity to promote a sectarian violence based
on irrational and mutually incompatible beliefs. In fact, while "religious"
factors undoubtedly played a role, largely because religious and political
motivations were then difficult to distinguish, these unfortunate conflicts
were as much about the rival territorial ambitions of the Habsburgs and
the Bourbons. This accounts for the otherwise puzzling fact that in these
"religious" wars, Catholics and Protestants could find themselves on the
same side. The idea of plural religions as codified sets of beliefs and specific practices that can exist independently of political considerations and
are capable of relegation to a "private sphere" was one of the end products
of this process of state building. Indeed, it is not a complete distortion to
reverse the received understanding of these wars and say that the formation of the modern state was their cause, and the modern notion of religion a consequence.[45]

For some, these new, objectively defined religions were entirely compatible with more traditional conceptions of religion as inner piety. Those
who held to the primacy of the interior dimension of religion could at the
same time uphold the importance of explicit belief and could also understand religion in those terms. For the first generation of Protestant Reformers and the English Puritans, for example, there was an intimate relationship between the inner and outer components of religion. As we have
seen, the major Reformers stressed both the importance of genuine piety
and the need for the profession of explicit beliefs.[46] For others, however, it
was possible to make a sharp distinction between the two dimensions of
religion, holding fast to an idea of a genuine interior faith while allowing
at the same time that positive religions could be understood as artificial

juridical constructions. The German philosopher and jurist Christian Thomasius (1655–1728) contended that religions, understood as competing creedal systems, were the combined creation of the clergy and the state. The rival claims of these humanly constructed religions were to be understood as arising out of particular historical circumstances rather than the inner piety to which they were distantly related. The insight that modern religions are human constructions thus dates back to the moment of their birth.[47] Yet even the most staunch advocates of a pietistic understanding of their own religion found themselves outsiders to the others. Granting this, they could only gesture toward some supposedly shared interior dimension. It was the more robust empirical features of these new religions that represented their most visible manifestation and, along with that, the basis for scientific study, for comparison, for philosophical analysis, and in the European case, for territorial division.

The inscribing of these religious boundaries onto the map of Europe, coinciding as it did with the beginning of Western colonial projects, was followed by the creation of what we now call world religions. These came into existence through the projection of the religious fragmentation of Western Christendom onto the rest of the world.[48] In much the same way that religious difference was now a feature of the geography of Europe, an analogous religious differentiation was thought to characterize the whole of the globe. These once embedded "religions" were similarly fashioned out of the beliefs and social practices of indigenous peoples. Initially, the West imagined there to be four such religions: Christianity, Judaism, "Mahometanism," and heathenism. This last category was to be extended in various ways to produce the full gamut of the world religions. One of the earliest attempts at a kind of comparative religion was William Turner's *The History of all Religions in the World* (1695), in which the rather vague "heathenism" is further subdivided into "ancient," "modern," and "diabolical." Turner describes these religions under two heads—their "theory" (that is, their beliefs) and their practices. Turner also distributes these religions geographically across the globe, in much the same way that the Christian religions had been distributed across the face of Europe (see figures 4 and 5). The "other religions" were thus constructed as inferior versions of the territorialized Christian religions of Europe.

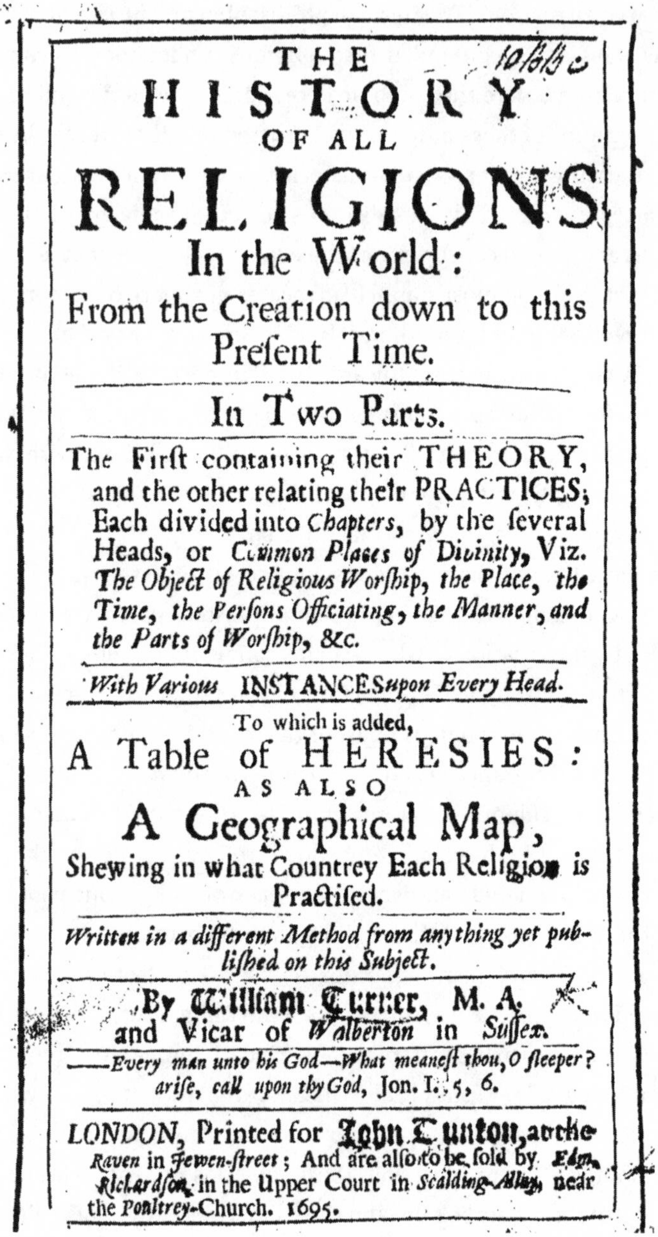

THE
HISTORY
OF ALL
RELIGIONS
In the World:
From the Creation down to this Present Time.

In Two Parts.

The First containing their THEORY, and the other relating their PRACTICES; Each divided into *Chapters*, by the several Heads, or *Common Places of Divinity*, Viz. *The Object of Religious Worship, the Place, the Time, the Persons Officiating, the Manner, and the Parts of Worship*, &c.

With Various INSTANCES *upon Every Head.*

To which is added,
A Table of HERESIES:
AS ALSO
A Geographical Map,
Shewing in what Countrey Each Religion is Practised.

Written in a different Method from any thing yet published on this Subject.

By William Turner, M. A.
and Vicar of *Walberton* in *Sussex.*

——*Every man unto his God*——*What meanest thou, O sleeper? arise, call upon thy God*, Jon. I. 5, 6.

LONDON, Printed for John Dunton, at the Raven in *Jewen-street*; And are also to be sold by *Edm. Richardson* in the Upper Court in *Scalding Alley*, near the *Poultrey*-Church. 1695.

Figure 4. Title page of The History of all Religions in the World (1695), one of the first works of "comparative religion," in which William Turner describes religions in terms of theory and practices.

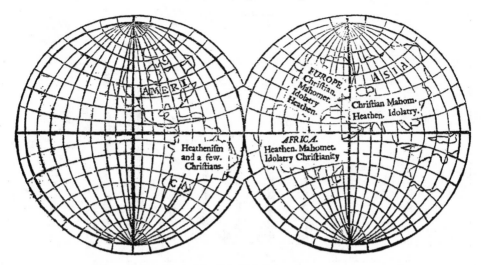

Figure 5. William Turner's Map of the "World Religions."

What are now often referred to as the "Eastern religions" would be differentiated much later, appearing in the nineteenth century and registering their birth with a new range of labels. The first recorded use of "Boudhism" was 1801, followed by "Hindooism" (1829), "Taouism" (1838), and "Confucianism" (1862) (see figure 6). By the middle of the nineteenth century these terms had secured their place in the English lexicon, and the putative objects to which they referred became permanent features of our understanding of the world. The precondition for the appearance of these plural "religions" was the emergence of a generic conception of religion, typically understood in terms of beliefs and practices, during the early modern period.

These new global religions made their own contribution to the modern idea of religion. Whereas the religions of Europe had been generated out of the objectification of interior states, the world religions were, in a sense, reverse engineered from the newly constructed Western religions. What were classified as religious practices thus provided the foundation for the inferences about religious beliefs. This proto-anthropology of religion necessarily brought with it the idea that religion was to be identified

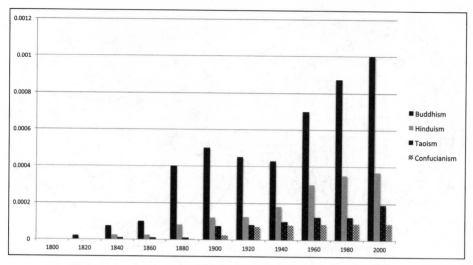

Figure 6. Frequency of the terms "Buddhism," "Hinduism," "Taoism," and "Confucianism" (including spelling variants) in English books, 1800–2000.

with what could be dispassionately observed. Since the external components of the religions could be established with much more certainly than the interior, the former became a proxy for the latter. There is a sense in which, with the proliferation of this understanding of religion, Western thinkers became religious outsiders to all traditions but their own.

True Religion Revisited

The appearance of discrete religions in Western Europe brought with it a new, and distinctively modern, problem: which religion is true? This problem was further exacerbated by the imaginative construction of non-European religions. When religion is understood in terms of an attitude of inner piety, the question of its rectitude relates to whether it is properly directed and motivated, or whether it strikes the right mean between the extremes of superstition and atheism. The patristic discussions of true religion that we considered earlier thus focused on worship and the proper object of worship. For this reason, true religion could not simply be equated with Christianity. With the first appearance of religions understood in terms of beliefs and practices, the question of truth could be directed to their propositional content. Furthermore, understood propositionally, only some religions—and possibly only one—will be true, while others will be necessarily false. Looking ahead, the possibility of this new kind of conflict between religions presages a potential conflict between religion and science, once both enterprises are understood as systems of practices and beliefs.

In the early modern period there were three possible responses to this conundrum. Most radical was that proposal that "all are refuted by all," as the French political philosopher Jean Bodin put it.[49] For the vast bulk of early modern thinkers, including Bodin himself, this relativistic option was a mere theoretical possibility and one that was to be resisted. A second solution was to contest the new reified religion and insist that true religion was still to be identified with piety. As we have seen, this position was characteristic of Christian Thomasius and of the aptly named Lutheran pietists, who emphasized the importance of the devotional life and high moral standards. Resistance to the idea that Christianity is a religion among others has been characteristic of a significant number of theologians and Christian communities since that time. The third response was to concede, or even embrace, the propositional nature of religion and seek ways to adjudicate between the rival claims of the religions thus understood. This option, or some version of it, was widely taken up, for it was consistent with the almost universal assumption that Christianity was the true religion, and with the conviction that its truth could be established by rational argument. Western science, increasingly understood as a unique embodiment of rationality, will come to play an important role in this process, either by confirming or contesting the truth claims of Christianity.

Robert Boyle addressed the problem of competing truth claims in an unpublished work on "The Diversity of Religions." Boyle is generally considered to be the father of modern chemistry, and is perhaps best known for the gas law that bears his name. Like many of his scientific contemporaries he also had an abiding interest in religious matters. Boyle points out that there are many nations in the world, and four great sects (Christians, Jews, Mohammedans, and pagans) each of which is divided into what he calls "several different systems of belief." Once we understand

religions as competing belief systems we are then confronted with a particular problem:

> no man of prudence or moderation will imagine that, surrounded by
> such a variety of opinions and warring sects, each with learned men
> amongst its followers, he is at all likely to embrace the one and only true
> religion, especially when everyone maintains that his religion is true, and
> all acknowledge that there is only one true one while some suspect that
> none is wholly true.[50]

Others expressed the conundrum in similar terms.[51] The truth of religion, thus understood, now came to hinge upon the vital question of how
one might adjudicate between competing claims of these "systems of belief." Clearly, this conception of the truth of religion is different from that
adopted by Augustine, for whom true religion had existed since the beginning of history, or Ficino, who held that any religion directed toward
God was true Christian religion, or even Swiss Reformer Ulrich Zwingli,
who thought that Christians were capable of exhibiting true and false religion. The seventeenth century witnesses the beginning of a quite different
understanding of true religion, and of how to establish the veracity of the
propositions that now constitute it.

One way to assess the authenticity of a historical religion was to trace
the lineage of the relevant beliefs and measure them against the original
and authoritative historical source of the religion. In the case of Christianity, this called for evidence of conformity between contemporary
forms of the Christian religion and the religion that Jesus had supposedly
taught and that had been passed down by his disciples. This was motivated
by the idea that there had been an original and ideal Christian religion in
the first century, the boundaries of which were defined by its doctrinal
content. In the seventeenth-century a clear indication that this criterion
is in play is the frequency with which we encounter such phrases as "the
true religion taught and established by the apostles."[52] English clergyman
and devotional writer Jeremy Taylor (1613–67) thus spoke of "the first Principles of the Christian Religion taught by Christ himself."[53] Taylor's contemporary the Puritan divine John Owen (1616–83) came from the oppo-

site side of theological spectrum but nonetheless spoke in similar terms of "the Religion taught by Jesus Christ, and contained in the Scripture."[54] Bishop Thomas Sprat (1635–1713), who, not coincidentally, was a key apologist for the Royal Society and the new experimental science, wrote that "Christ himself understood better the Interest and Power of his own Religion. He knew, that the design of his coming into the World, was . . . to introduce a rational, moral, spiritual Doctrine, and a plain, unaffected, saving way of teaching it."[55] In sentiments such as this we encounter the idea that Jesus had professed a propositional religion. This is not to say that personal piety became unimportant for these individuals. Indeed for many it remained a central aspect of their understanding of the Christian life. Yet the traditional emphasis on piety was now accompanied by a new understanding of Christianity as a propositional religion that had been instituted by Christ. This is a novel development, since traditionally God was thought to have revealed himself in Christ, and not a religion. This line of argument is not especially vulnerable to philosophical or scientific objections, but historical critical methods can render judgment upon it.

Historical arguments might help settle the question of which Christian confession could lay claim to truth, but the rise of the concept of religion also led to the possibility of formulating the more radical question of whether Christianity itself, in any of its confessional forms, was true. Accordingly, in addition to these arguments about the historical authenticity of particular doctrinal traditions there developed a range of arguments that made appeals to the rationality of Christian beliefs. The seventeenth-century religious literature abounds with titles offering to discuss the "grounds and reasons," "evidences," "proofs," and "vindications" of the Christian religion, along with "impartial comparisons" and "surveys" of competing creeds.[56] Robert Boyle, who had so starkly set out the predicament of the seventeenth-century religious believer, thus proposed that one should only choose a religion "having maturely weighed the grounds of it, and compar'd it, and the arguments for it, with the plea's that occur on the behalf of each of the grand Religions that differ from it."[57] Not surprisingly, perhaps, these comparative exercises usually (although not invariably) arrived at the conclusion that belief in the doctrines of the Christian religion was best supported by the comparative evidence.

Richard Baxter's *Reasons of the Christian Religion* (1667) offers a typical example. Baxter, a prominent Puritan writer and nonconformist minister, promised the reader on the title page that the book will both prove "by natural evidence the existence of God" and "by evidence supernatural and natural the certain truth of the Christian belief."[58] The idea that a religion can be rationally justified is dependent on this new understanding of religion. For those who subscribed to this ideal, the perfect religion would be a body of propositions, firmly established by ironclad logical demonstrations. The inhabitants of the fabled kingdom of Macaria—a Puritan Utopia described in an anonymous tract published in 1641—possessed just such a creed: "Their Religion consists not in taking notice of severall opinions and sects, but is made up of infallible tenets, which may be proved by invincible arguments, and such as will abide the grand test of extreme dispute."[59] These approaches to religion were to usher in what John Henry Newman was to disdainfully designate, from the vantage point of the nineteenth century, "the Age of Evidences."[60]

Accompanying this emphasis on rational proofs was a modified understanding of faith and belief. As we have seen, belief once entailed trust in a person, and faith had been an infused virtue. Now belief could be described as the act of giving intellectual assent to propositions. Isaac Barrow (1630-77)—the brilliant theologian, classicist, and mathematician who preceded Isaac Newton in the Lucasian Chair at Cambridge—thus declared that "the proper object of faith is . . . some proposition." This might be extended to "some system of propositions" as in the specific case of "all the propositions taught in the true religion as so." As for faith in a person (*fiducia*), this was in effect reducible to giving credence (*assensus*) to propositions about that person: "to believe on a person, or thing is only a short expression (figuratively) denoting the being persuaded of the truth of some proposition relating in one way, or other, to that person."[61] In the idiom of modern analytic philosophers of religion, "belief in" is to be equated with "belief that."

Such a conception is rather different from the patristic idea of belief as related to trust and obedience. It also differs from the position of Thomas Aquinas who, in the very first question on faith in the *Summa theologiae*, inquires after its proper object. He concludes that the proper object of faith is not some proposition, but a thing—the "First Truth" or God him-self.[62] Aquinas clearly allows that propositional truths—the articles of faith—are important, but he insists that the grounds for assenting to the truth of these articles lies in the trustworthiness of God rather than, say, the testimony of human reason.[63] As for belief *in God,* as opposed to be-liefs *about God,* Aquinas follows Augustine in holding that the former is not reducible to the latter, since our confidence in the truth of beliefs about God ultimately depends upon our attitude of trust toward the divine source of those revealed truths.[64] The articles of faith, moreover, are in-evitably incomplete and partial: "The things of faith," says Aquinas, "are not proposed in themselves but by certain words and likenesses which fall short of expressing or representing them; consequently they are said to be known as through a mirror in a dark manner."[65] To some degree, then, all assent to propositions was necessarily implicit, given the lack of precision in the verbal formulas that captured theological truths.

Something of this older association of belief with trust is maintained in Francis Bacon's early discussions of heresy. In his *Meditationes Sacrae* (1597), Bacon suggests that atheism and heresy are not to be understood primarily as a lack of appropriate propositional belief in the one case and of subscribing to false propositional beliefs in the other. Rather, both arise out of a lack of trust—they "rebel and mutiny against the power of God; not trusting to his word, which reveals his will."[66] Certainly, atheism and heresy involve false beliefs, but those beliefs are the consequence of a more fundamental moral failing. In this respect, Bacon's understanding of heresy is akin to that of Dutch humanist Erasmus of Rotterdam (1466–1536) who had earlier proposed four necessary conditions for heresy: the perversion of doctrine, stubborn persistence in error, the attempt to attain personal advantage, and the presence of malice.[67] The strong opprobrium attached to heresy, often puzzling to those living in a post-Enlightenment world, can thus partly be attributed to this moral dimension.

By way of contrast, for a number of influential English Protestant think-ers in the second half of the seventeenth century, faith and belief were to be reduced to a cognitive act rather than a relational virtue. What made faith valid was the inherent rationality of the propositions toward which it was directed. In his aptly titled *Rational Account of the Grounds of the Prot-*

estant Religion (1664), theologian Edward Stillingfleet declared that faith
is "a rational and discursive act of the mind . . . an assent upon evidence, or
reason inducing the mind to assent." The stance of John Locke—the philosopher most closely associated with the new experimental science—
also exemplifies this new approach to faith. In spite of his disagreements
with Stillingfleet over key theological issues, Locke agreed with him that
"faith" was to be understood as "giving assent to propositions," and in the
specific case of religious faith, as "an assent founded on the highest reason."[68] Locke explicitly repudiated the notion of faith as trust (fiducia), because in his view it led to implicit faith and stifled inquiry: "whilst some
(and those the most) taking things upon trust, misemploy their power of
assent, by lazily enslaving their minds to the dictates and dominion of
others in doctrines, which it is their duty carefully to examine, and not
blindly, with an implicit faith, to swallow."[69] Implicit faith, or faith in the
sense of fiducia, was merely "assent without enquiry."[70] Locke also speaks
in his Letters Concerning Toleration, of "the true religion," which is equated
with "the Christian religion."[71]

It is at this point that the new science comes to play an important role,
because the natural history and natural philosophy of the seventeenth
century are charged with the mission of providing some of the general
warrants needed by the new propositional religion. In doing so, they make
their own contributions to the reification of religion—that is to say, religion understood as a system of beliefs requiring some degree of rational support—while at the same time consolidating their own status as
religiously useful activities. The shape that modern science takes at this
time is thus partly determined by its interactions with the newly ideated
"religion." As a consequence of this, the means by which scientific and
religious knowledge are thought to be attained will converge, so that to
some degree science and religion will come to share a common epistemic
basis.[72] But the establishment of this common cognitive ground that made
it possible for the new science to support the new religion gave hostages to
fortune, for in this initially positive relationship between science and religion lay the seeds of what would become a more conflicted future.

Science and the Age of Evidences

The new kind of partnership between science and religion, as should now be evident, was predicated on their both becoming sets of practices and beliefs. The most obvious way in which the early modern sciences of nature contributed to the new conception of religion was through the provision of rational evidences for its propositional content. For many of its practitioners, the study of nature necessarily included reference to principles of design, and the motif of the divinely authored "book of nature" became even more prevalent than it had been during the Middle Ages. To be sure, nature was no longer to be "read" for its deeper theological and moral meanings, yet it was widely believed that systematic investigation of its underlying principles would point to the purposeful design of a wise and powerful Deity. The cosmos, in other words, might not offer analogies of the Trinity or recapitulations of the Christian narrative of redemption, but it was thought to provide a limited number of premises from which some basic truths about God could be inferred. Isaac Newton spoke for a number of his scientific contemporaries when he declared, in the celebrated General Scholium appended to the 1715 edition of the *Principia*, that "this most elegant system of the sun, planets, and comets could not have arisen without the design and dominion of an intelligent and powerful being." Indeed for Newton, natural philosophy could not be wholly distinct from theology, for he went on to say that discourse about God "is certainly part of natural philosophy."[73]

It is sometimes assumed that Newton's inclusion of God in the subject matter of natural philosophy was a restatement of a traditional view, a harking back to previous centuries during which there was a more intimate connection between what we now call science and religion.[74] But in fact this was a quite novel claim, portending the present authority of science rather than nostalgia for the medieval past. From the twelfth century onward, the demarcation between natural philosophy and theology had been scrupulously upheld, and Aristotle's understanding of the discrete subject matters and methods of the two disciplines was accepted both in principle and practice. Institutionally, this was reflected in the way in which the universities were organized, with the different curricula and goals of the faculty of arts and the higher faculty of theology. Although inevitably there were occasional demarcation disputes, in theory, those trained in natural philosophy in the arts faculty would not have presumed to speak on theological matters. However, the seventeenth century sees a blurring of the lines between natural philosophy and theology, and the emergence of new disciplinary combinations and vocational identities.

The early modern category of "physico-theology" nicely captures the new overlap of interests, and the very name embodies the idea that physics (or natural philosophy) might profitably be combined with theology. This amalgamation was one of two key hybrid disciplines that emerged during the period of the scientific revolution, the other being "physico-mathematics," which, as the name suggests, saw the fruitful combination of natural philosophy with the third of Aristotle's theoretical sciences — mathematics.[75] Again, this new combination was in breach of Aristotle's insistence that the theoretical sciences be kept apart. Historians of science have typically understood the significance of the new combination of mathematics and physics in terms of the new kind of science that it yielded, culminating in Newton's magnum opus, the *Principia Mathematica* (The Mathematical Principles of Natural Philosophy, 1687) — a work whose very title expresses the new combination of disciplines. Yet the breakdown of the traditional disciplinary divisions is significant for another reason, for it signals an abandonment of the idea of the three theoretical sciences as stages in a process of personal formation in which the mind gradually progresses away from material things, through the more abstract objects of mathematics, to theology. As for physico-theology, it reflects the conviction that what natural philosophers study is God's activity, both his direct causal activity and his design of the creatures. This is because divine causation is now considered to be more or less identical to natural causation, and because the theological significance of the creatures is no longer understood allegorically, but in terms of the way in which they have been designed.

Natural philosophers thus come to assume a new vocational identity during this period, becoming what historian Amos Funkenstein has called "secular theologians" — individuals with no formal training in theology who nonetheless consider themselves capable of advancing theo-

logical arguments.[76] Johannes Kepler (1571–1630), the gifted German astronomer who first provided accurate mathematical descriptions of the
planetary orbits, thus wrote in a letter to his old teacher Michael Maestlin
that "I wished to be a theologian; for a long time I was troubled, but now
see how God is also praised through my work in astronomy."[77] Later he
would affirm that the whole world was "the temple of God" and "the book
of Nature." To contemplate nature was thus to engage in "true worship
to honor God, to venerate him to wonder at him."[78] Robert Boyle also argued that the world was the "temple" of God and that this gave warrant to
the priestly vocation of investigators of the natural world. Boyle observed
that "philosophers of almost all religions have been, by the contemplation
of the world, moved to consider it under the notion of a temple."[79] Accordingly, the contemplation of nature was "the first act of religion, and
equally obliging in all religions." Natural philosophy was nothing less than
the "philosophical worship of God." He went on to suggest that "discovering to others the perfections of God displayed in the creatures is a more
acceptable act of religion, than the burning of sacrifices or perfumes upon
his altars."[80] For Boyle, ordination in this "priesthood" called for a knowledge of the operations of the natural world. The true priestly vocation,
in short, could be followed only by those familiar with the workings of
nature.[81]

As Boyle implied, exponents of the new philosophy could claim to have
a unique expertise in offering new evidences for the truth of the Christian
religion. Newton's conclusions about a designing Deity, for example, are
appended to some five hundred pages of the most recondite mathematical
reasoning, the full significance of which would have remained obscure to
all but a handful of his contemporaries. Robert Hooke, pioneer and popu-
larizer of microscopy and author of the incomparable *Micrographia* (1665),
observed that the more we magnify objects, "the more we discover the im-
perfections of our senses, and the Omnipotency and Infinite perfections of
the great Creatour."[82] Again, the evidences offered in Hooke's *Micrographia*
were mediated by an individual who had spent countless hours designing and constructing a sophisticated optical instrument, mastering its
use (which was then no small matter), and then producing the beautifully
detailed drawings for which the work has become justly famous. Robert

Boyle also believed that the experimental philosophy would provide "new Weapons for the defence of our ancientest *Creed*."[83] The novel methods of investigation adopted by the new generation of natural philosophers had uncovered whole new territories in the vastness of space and the narrow confines of the microscopic world. As Boyle expressed it, only the expert investigator can access the "more penetrating indagations of the abstrusities of nature, and the more unobvious properties of things."[84] These novel features of the natural world, hitherto unimagined by generations of scholars steeped in Aristotelian teachings, provided new ammunition for theistic arguments and thus established the privileged theological authority of experimental philosophers. The deliverances of specialist natural philosophy could thus be contrasted favorably with what Thomas Sprat disparagingly designated "the blind applauses of the ignorant."[85] And it was not just any natural philosophy that could play this role. This was claimed to be a distinctive feature of the *new* scientific approaches. Fortuitously, then, at the precise historical moment when the newly conceived "religion" found itself in need of rational evidences, the new natural philosophy was able to provide them.

Natural philosophers were also highly attuned to the different kinds of evidences that they could furnish for establishing the truth of the Christian religion. At this time religious evidences were understood to be either "internal" or "external." The former pertained to the inherent rationality of the religious beliefs and typically included traditional arguments for the existence of God; the latter were to do with the credibility of those who promulgated particular religious doctrines. If a teacher of particular religious doctrines were a miracle worker or a prophet (and provided in the latter case that those prophecies were subsequently fulfilled), this would represent "external" evidence for the doctrines that they had taught.[86] The capacity to work miracles or to predict the future, in other words, provided indirect evidence of trustworthiness. If internal evidences provided support for religion in general, external evidences provided arguments for one religion in particular.

As we have seen, experimental philosophy was widely acknowledged to provide internal evidence for God's existence, primarily by educing the many instances of design in nature. But experimental philosophy also supported specific Christian truths by attesting to the likelihood of miracle claims associated with the promulgation of the Christian religion. Robert Boyle maintained that "Experimental Philosophy does also dispose the minds of its cultivators to receive due impressions from such proofs as Miracles do, as well as other Topicks, afford the Christian Religion." Elsewhere he was to contend that natural philosophers, above all others, have the ability to examine miracles, prophecies and other proofs "that are alleged to evince a real religion."[87] Bishop Sprat, in his defense of the activities of the early Royal Society, agreed that natural philosophers were uniquely qualified to judge the authenticity of "prophetical visions" and "extraordinary events."[88]

It is tempting to frame these efforts as apologetic exercises on the part of pious scientists—the lending to religion of the authoritative voice of science. But if this was an apologetic strategy, it was as much about supporting science as Christianity. Science—or, to speak more accurately, the new experimental natural philosophy—did not yet speak with an authoritative voice and, as will become evident in the next chapter, during this formative period it was the new philosophy that needed to establish its epistemic credentials, its religious respectability, and its social utility. That said, these friendly overtures on the part of the natural philosophers had the unforeseen consequence of raising the profile of certain elements of Christian religion—namely, those propositions to which natural philosophy could lend its support. In this manner, the attentions of the natural philosophers further promoted the reification of religion, and shifted attention away from the formative dimension that had once been central to both enterprises.

These developments accompany the emergence of a number of modern disciplines that are now biased toward questions of knowledge and its justification. Thus, in these physico-theological exercises we witness the birth of natural theology, understood as the provision of supporting arguments for theological doctrines, based on reason alone. Early modern talk of contrivance and design in nature was not simply the continuation of a medieval tradition of natural theology with a better and more objectively established data set. Rather, it was the institution of a new approach to nature that was at once more modest and more ambitious than the theologies of nature that preceded it: more modest, because it sought to demonstrate the truth of a relatively restricted range of ideas about God; more ambitious, because it aimed to do so on the basis of a putatively neutral reason alone, and making fewer assumptions about shared, preexisting theological commitments. Part of the novelty of this natural theology came from the fact that the parent discipline of theology itself was in the process of being reconceptualized. Given the changes taking place to the understanding of *religio* and *scientia,* it could hardly have been otherwise. The "science" of theology, as we have seen, had been understood as a faculty or habit, and the study of theology called for a mental and moral habituation. This sense of the term was also captured by the English word "divinity," which can bear the double meaning of "the quality of being divine or godly" and "the science that deals with God and his attributes." Increasingly, the meanings of both "divinity" and "theology" shifted to the latter sense. Finally, we can say that natural philosophy became increasingly disengaged from philosophy, which was then still to some extent understood in terms of the classical conceptions of the philosophical life. The transformations under consideration here thus have significant implications for our understanding of modern philosophy, which in the Anglo-American tradition is imagined to be primarily about epistemology.

Related to this is the modern idea of knowledge as "justified true belief," along with the assumption that what constitutes an appropriate justification should be drawn from the sphere of natural philosophy. The most extreme exemplification of this tendency came with the demands of the logical positivists of the last century that aesthetic, moral, and religious claims be subjected to the verification principle that was then thought to characterize scientific knowledge. Strict adherents of this criterion insisted that no proposition was meaningful unless it was capable, in principle, of some kind of empirical confirmation. While it is now widely recognized that this was a principle that few forms of knowledge could satisfy—including many doctrines of the natural sciences—the notion that religion consists of factual claims that should be subject to scientific confirmation is common enough in present-day science-and-religion discussions and is accepted not only by scientific critics of religion, but by many religious apologists as well.

It seems hardly necessary to point out that developments of this kind, while initially pursued in support of the doctrines of Christianity, were to render "the Christian religion" susceptible to new avenues of criticism that focused on the rationality of its propositional contents. As historian Michael J. Buckley has shown, the complexion of Enlightenment atheism was directly informed by the particular version of Christian religion that it sought to deny.[89] To the extent that Christianity was now understood to be "a religion," constituted by beliefs that were supported by certain kinds of evidences, it invited a denial of those beliefs and their grounds, and thus of the whole enterprise for which those beliefs had come to stand. In England, Christian apologists were confident that positive relations between natural philosophy and theology would endure, and for a time they were correct. The natural sciences gained considerable social legitimacy through their sharing of intellectual territory with religion. Whether religion was a long-term beneficiary of this positive relationship is more doubtful. Arguably, religion thus construed became vulnerable to particular lines of criticism—hence the aphorism that no one doubted the existence of God until the Boyle lecturers undertook to prove it. Yet it is clear that in the middle of the eighteenth century the religious hypothesis about the origins and design of the world represented a kind of inference to the best explanation (although David Hume and Pierre Bayle had already mounted challenges against such an inference). This stance was possible only because of the current state of natural history. The advent of Darwinism rendered this less plausible, although more damage to propositional belief was wrought by a new historical consciousness and the advent of biblical criticism.

::::

It is worth observing, by way of conclusion, that with the rise to prominence of a new conception of religion the older notion of religion as piety did not disappear. As we have already seen, Christian Thomasius railed against "brain religion" that placed propositional belief ahead of love of God and neighbor.[90] In the eighteenth century, John Wesley was likewise to insist that the proper objects of faith were not propositions, but God and Christ. Faith, he argued, was not "a speculative, rational thing, a cold lifeless assent, a train of ideas in the head; but also a disposition in the heart."[91] From Puritanism, to Methodism, to Lutheran Pietism, to Schleiermacher's "feeling of absolute dependence," to Kierkegaard's "attack upon Christendom," there have been strong reactions against over-rationalized and propositionalized versions of Christian faith. More recently, there have been comparable rejections of the idea of Christianity as a religion. In the twentieth century Neoorthodox theologian Karl Barth famously declared "religion" to be unbelief. His younger contemporary Dietrich Bonhoeffer advocated a "religionless Christianity." Adherents of other so-called religions are at times equally adamant that their commitments and ways of life should not be so classified.[92]

So it is not the case that the understanding of religions as sets of beliefs and practices totally displaces the earlier notion. The point is rather that from this time on religion and religions *can* be understood in terms of beliefs and practices that are empirically available for comparison and analysis. Religion now exists concretely as something that can serve as an explanation for historical events and which in turn can be "explained" by various social sciences. Religion thus understood can be identified as a "cause" of violence and terrorism and is given a prominent role in interpretations of world events, particularly those that draw upon the idea of a "clash of civilizations" made popular by political scientists such as Samuel P. Huntington. Religion also offers itself as a phenomenon that can be accounted for by cognitive scientists or evolutionary psychologists such as David Sloan Wilson, often on the assumption that the persistence of obviously false beliefs cries out for such explanations. Finally, it is religion thus conceived that subjects its beliefs to confirmation or disconfirmation by the modern disciplines of philosophy and science.

[5]

Utility and Progress

What does it profit a man if he gain the whole world, but lose his own soul?
—Mark 8:36

The greatest inventions were produced in the times of ignorance, as the use of the compass, gunpowder, and printing, and by the dullest nation, as the Germans.
—Jonathan Swift, "Thoughts on Various Subjects"[1]

Again there is another great and powerful cause why the sciences have made but little progress; which is this. It is not possible to run a course aright when the goal itself has not been rightly placed.
—Francis Bacon, *New Organon*, I, §81

In an essay contributed to the *Edinburgh Review* in July 1837, the poet, politician, and historian Thomas Babington Macaulay offered this perceptive assessment of the tenor of Francis Bacon's new philosophy, and of how it differed from what came before: "Two words form the key of the Baconian doctrine, Utility and Progress. The ancient philosophy disdained to be useful, and was content to be stationary. It dealt largely in theories of moral perfection. . . . It could not condescend to the humble office of ministering to the comfort of human beings." Macaulay went on to explain that natural philosophy had marked time since its inauguration by the ancients, owing to its preoccupation with ethical formation. The admixture of ancient philosophy with Catholic scholasticism in the Middle Ages had done nothing to change this. But then, in the wake of the reformation of religion, "Bacon appeared." He was armed with a new philosophy. Its law was progress, its object "the good of mankind."[2]

In recent decades, Macaulay's approach to history has come to be regarded as quintessentially "whiggish," meaning that it exemplifies an approach in which history is understood in terms of an inexorable march toward progress. Past events are judged in terms of the extent to which they contribute to those things that are judged to be worthwhile in the present. Macaulay gives the game away when he unabashedly announces in the opening paragraphs of his *History of England,* that "the history of our country during the last hundred and sixty years is eminently the history of physical, of moral, and of intellectual improvement."[3] This emphasis on inevitable progress is shared with other nineteenth-century visions of history. Karl Marx (1818–83) thus contended that humanity is moving ineluctably toward the classless society, which will represent a kind of ideal end state. A similar idea is found in the famous "three stages" of the positivist philosopher Auguste Comte (1798–1857), who argued that "each branch of our knowledge . . . passes successively through three different theoretical conditions: the theological, or fictitious; the metaphysical, or abstract; and the scientific, or positive."[4] Needless to say, perhaps, whiggish history is at present deeply unfashionable, particularly among professional historians of science. But it continues to inform popular histories of science, secular humanist manifestos masquerading as history, and some American neoconservative interpretations of world affairs.[5]

Leaving aside for the moment Macaulay's reliance on a now unfashionable theory of history (and his rather jaundiced view of Bacon's moral character), it can be said that his observations on the general significance of Bacon's ideas are deeply insightful. He sees clearly that Bacon's novelty is to be understood against the background of a traditional conception of natural philosophy that focused on moral perfectibility. In light of this, Bacon's contribution is rightly identified as that of providing new goals for natural philosophy—goals that orient it toward a more general human welfare. In addition, Macaulay singles out utility and progress as key elements of Bacon's thought. These two notions, along with the distinctively Christian ideas of charity and dominion over nature, will provide the main focus of this chapter.

The idea of historical progress is usually thought to have originated in the seventeenth century. My main argument in this chapter will be that changes to our conceptions of science and religion at this time were intimately related to the emergence of a new understanding of progress. One significant factor in this development is the way in which early modern natural philosophy self-consciously locates itself within a Christian understanding of history—as part of a preordained reformation of knowledge, as a prelude to the end of the world, as a providential and partial restoration to the human race of knowledge lost at the Fall, or some element of all of these. It has often been said that notions of progress represent a secularized Christian eschatology, and much has already been written about this. But my primary focus here will be the way in which the idea of progress—or perhaps more correctly, a distinctively modern idea of progress—appears for the first time at this juncture in history, and how it is related to the changes to *scientia* and *religio* that we discussed in the previous chapter. Progress is now understood as arising out of the cumulative contributions of numerous individuals to a body of knowledge. As we will see, such a conception of progress is possible only when "science" and "religion" are entities that exist outside and independent of individual human persons, and of which doctrines or propositional content are a central component.

As we shall also see in this chapter, other concepts are implicated in this new idea of progress. Most important for our purposes are the ideas of usefulness, charity, and dominion. In each of these we see the beginnings of a decisive shift away from the formative and personal to the progressive and objective. One of the consequences of these transitions is the possibility of a new normative claim that science is progressive—at least in the way in which progress is now understood—in a way that religion is not. Once the categories of science and religion become established in this way, the new understanding of progress lays the foundations for a range of common claims about the superiority of science to religion, and about the secular and scientific West as the apex of cultural evolution.

Two Ideas of Progress

In the first chapter we saw how Thomas Aquinas has an idea of the growth of "science" (*scientia*) as the development of a particular human capacity: "science can increase in itself by addition; thus when anyone learns several conclusions of geometry, the same specific habit of science increases in that man."[6] In fact, most of Aquinas's references to progress (*profectus*) in the *Summa theologiae* relate to personal progress toward virtue.[7] This idea of progress sits within the broader context of a general teleology, common to a number of the classical schools and to medieval theology, in which the goal-oriented entities move naturally toward their ends. It is this movement that counts as progress. As the Roman emperor and Stoic philosopher Marcus Aurelius put it: "That which is the purpose of each thing's construction and the destination of that construction is the destination of its progress; the destination of its progress is that in which its goal lies; where its goal lies is where each thing's advantage and good lie."[8] Human progress is thus understood to be the natural movement of the individual toward the goals of wisdom and virtue. The natural world provides the context in which progress takes place, being ordered in such as way as to provide material benefits for its human tenants, but also sufficient obstacles to encourage their moral and intellectual development.

Without a teleological framework, this personalized understanding of progress becomes redundant. In the specific case of the virtue *scientia*, medieval thinkers tended to think of its increase in terms of the augmentation of personal capacities. Once *scientia* becomes a reified entity, a "something" that exists objectively, progress comes to mean something quite different, namely, the cumulative addition to, and incremental improvement of, an external body of knowledge. It is often said that the idea of progress first appears in the West with the advent of modernity, and indeed that it is one of the hallmarks of modernity. But another way of describing this development would be to say that the idea of progress is relocated from the sphere of the individual into the historical realm—a realm that, incidentally, is inhabited by these newly reified notions "science" and "religion." On this new understanding, progress will no longer be trammeled within the relatively short life span of the individual, and the sciences become identified with a quite different conception of human progress, exemplified in a long term and open-ended historical development. Nature will also be given a new role in the process. Not simply the mere backdrop against which human advancement takes shape, it will itself become the subject of improvement.

The French mathematician, theologian, and natural philosopher Blaise Pascal (1623–62) provides us with an explicit connection between these two notions of progress. At this pivotal moment of historical transition, Pascal lit upon the arresting image of a whole sequence of men being substituted for the growth of the single individual. "Thence it is," he says, "that by an especial prerogative, not only does each man advance from day to day in the sciences, but all mankind together make continual progress in proportion as the world grows older, since the same thing happens in the succession of men as in the different ages of individuals. So that the whole succession of men, during the course of many ages, should be considered as a single man who subsists forever and learns continually, whence we see with what injustice we respect antiquity in philosophers."[9] Francis Bacon had earlier made a related point, suggesting that just as we look for greater knowledge of human things in the old than the young, so we ought to expect much more from the modern age than antiquity.[10] For both Pascal and Bacon this confidence in the progression of human knowledge is directly related to a new kind of natural philosophy that, to use Pascal's words, "consists in experiments and not demonstration."[11]

While this sounds very much like a simple advocacy of a new experimental method, it is important to understand that these assertions were directed as much against the traditional *goals* of the natural philosophy as its *methods*. It was not simply a matter of replacing logical demonstration with experiment, but in coming to understand that philosophy should be concerned less with personal edification and more with contributing to a common storehouse of knowledge. Bacon made the point that even the most earnest seeker after truth, following the traditional understanding of the goals of the sciences, "will propose to himself such a kind of truth as shall yield satisfaction to the mind and understanding in rendering causes for things long since discovered, and not the truth which shall lead to new assurance of works and new light of axioms."[12] Satisfaction to the mind and understanding is indeed progress of a kind, but Bacon urges upon his readers a quite different notion of the advancement of learning. A focus on matters of fact, discovered by experimentation or accumulated in the gathering of what he calls "natural histories," suggests the possibility of building up a repository of contingent truths that will always be capable of further augmentation. Already, Bacon says hyperbolically, we have a magazine of "infinite experiments and observations." Again, then, perfectibility and progress do not relate to the individual soul, but to products of knowledge. The "perfection of the sciences," as Bacon put it, will come "not from the swiftness or ability of any one inquirer, but from a succession [of them]."[13]

It is worth noting that for critics of the new experimental approach, the accumulation of contingent facts did not count as true science or philosophy at all, because it did not involve engagement with eternal and unchanging truths. To put it another way, this new cumulative approach did not give rise to demonstrative knowledge—knowledge that was necessarily true in logical terms—nor did it involve the mental rehearsal of those logical demonstrations that helped cultivate an appropriate scientific *habitus* in the knowing subject. One Aristotelian critic of experimental philosophy contended that experimental knowledge was "utterly Incompetent or Unable to beget Science" because it is "meerly Historical, and Narrative of Particular Observations; from which to deduce Universal Conclusions is against plain Logick, and Common Sense."[14] But in time, even those who found the new philosophy congenial were to concede this. John Locke was of the view that "rational and regular experiments" give us a better idea of the nature of things than any other approach. But he acknowledged that "this is but judgment and opinion, not knowledge and certainty." A natural philosophy based on experiments, he readily admitted, "is not capable of being made a science."[15] Yet, he implied, so much the worse for science, traditionally conceived. Accordingly, this long-term cumulative approach was recognized as requiring a quite new understanding of natural philosophy, and one that meant it was no longer possible simply to align it with "science" as it had been traditionally understood.

It is more than a happy coincidence that the idea of an accumulation of knowledge outside of the minds of individual knowers emerged just after the appearance of the new technology of the printed page, which made such a conception of knowledge possible in practical terms. In the Middle Ages, as we have repeatedly noted, *scientia* was an intellectual virtue and hence existed within individuals. Insofar as there was a recognition of the possibility of knowledge independent of individual knowers, this was limited to a small number of books written by recognized authorities.[16] These were housed in modestly proportioned libraries that could be thought of more as aids to the inculcation of the habits of learning than as storehouses of knowledge. The great explosion of printed sources in the sixteenth and seventeenth centuries made possible an accumulation of knowledge that vastly surpassed the capacity of individual minds. Bacon's "infinite experiments and observations" represented a body of facts that could only be stored in something like the print medium.

Modern speculations about the identity of "the last man to know everything," idle though they may seem, do point to this juncture in history when mastery of the sum of learning was beginning to look impossible. Candidates for this honor are typically those who lived during the early modern period: Joseph Scaliger (1540–1609), Francis Bacon (1561–1626), John Milton (1608–74), Athanasius Kircher (1601–80), Gottfried Leibniz (1646–1716). The epithet "Renaissance man" still carries with it the idea of comprehensive knowledge of a variety of fields.

The growing disproportion between what could be comprehended within a single mind and the rapidly increasing catalogs of printed material was often commented upon. Leibniz, despite his reputation as a polymath, bemoaned "the horrible mass of books which keeps on growing." The English poet Alexander Pope, likewise, complained of the "deluge of authors" inundating the intellectual landscape. This disquiet was not simply on account of what we now call "information overload," but also because it brought with it a novel conception of learning that was inconsistent with the older understanding of the relationship between knowledge and the intellectual virtues.[17] Pope's friend and fellow satirist Jonathan Swift maintained that "if books and laws continue to increase as they have done for fifty years past, I am in some concern for future ages, how any man will be learned."[18] His concern was not for the state of knowledge per se, but for the knower. To a degree, a similar anxiety informs present-day concerns about the impact of new technologies on our habits of mind: does easy and instantaneous access to information have a detrimental effect on attention spans, for example?[19] While these early modern concerns may at first blush seem rather exotic, they still have a contemporary resonance. To return to the point, once knowledge is relocated into new physical media, improvement and progress will pertain not to individual minds, but to a common storehouse of knowledge to which many minds will contribute and from which they will draw.

Since the sum of what could be known exceeded what could be contained within even the most capacious mind, it no longer made sense to justify the acquisition of knowledge primarily in terms of the role that it plays in intellectual and moral formation. A finite body of knowledge would do that just as well. So the usefulness of this ever-increasing pool of information was related instead to a general improvement in human welfare, understood primarily in material terms. Such a view necessitated a major revision of the traditional goals of natural philosophy, and of the end or use of the sciences. As we saw in the previous chapters, Bacon would speak of a natural philosophy directed toward "the uses of human life." Descartes spoke similarly of "the invention of innumerable devices which would facilitate our enjoyment of the fruits of the earth and all the goods we find there, but also, and most importantly, for the maintenance of health, which is undoubtedly the chief good and the foundation of all the other goods in this life."[20]

The Utility of Knowledge

Whereas learning had once been primarily for the edification of the individual subject, now the expectation was that learning should be useful for the commonweal. In our own age, which sees enormous investment in the natural sciences, and particularly those thought to yield economic benefits, it is hard to imagine that there was ever any question about the superiority of knowledge that yields practical and useful applications. But we get some sense of the novelty of the connection between learning and practical utility by looking at early modern controversies over this very question.

One indication of the contested nature of these novel conceptions of progress and usefulness comes from Jonathan Swift's savage satire on the experimental science of the Royal Society, set out in *Gulliver's Travels*. On his voyage to Laputa, Gulliver pays a visit to the grand academy of Lagado, where he encounters a cast of odd characters engaged in a variety of implausible projects—extracting sunbeams from cucumbers, softening marble for use in pillows, procuring colored cloth from spiderwebs, building houses from the roof downward, and many other schemes that Gulliver innocently admits he is "not skilful enough to comprehend."[21] Swift's derisory observations on the vanity of the new experimental science probably represent the best-known instance of attacks on the utility of the sciences, but these remarks were neither unprecedented nor uncommon. In the mid-seventeenth century James Harrington (1611–77) had written of the Oxford experimental club (a precursor to the Royal Society) that they were "good at two things, at diminishing a Commonwealth, and multiplying a Louse." The Oxford divine Robert South (1634–1716), also unimpressed by the microscopic investigations of Robert Hooke, Henry Power, and others, wittily observed that members of Royal Society "can admire nothing except fleas, lice, and themselves."[22] Even Charles II, despite being patron of the society, was said to have laughed mightily at the exploits the early fellows, who seemed to be preoccupied with such frivolous activities as "the weighing of air."[23] That these criticisms had a significant public resonance can be gauged by the success of Thomas Shadwell's play *The Virtuoso*, which opened in the Dorset Gardens Theatre in 1676 and was revived many times over the course of the next twenty years.[24] Shadwell cleverly personified the ambitions of the fledgling Royal Society in the enduring character of "Sir Nicholas Gimcrack" who, anticipating Swift's academicians, was engaged in a number of improbable experimental proj-

ects, most of which can be related to contemporary reports of experiments
conducted by the Royal Society.[25] Gimcrack took a special pride in the fact
that "we virtuosos never find out anything of use."[26]

These satirical observations are illustrative of the fact that the social
status of the new natural philosophy was by no means as secure as we
sometimes assume. Champions of the experimental approach found it
necessary to appeal to factors extrinsic to science in order to establish its
social utility. Part of their strategy was to appeal to religion, the legitimacy
of which was at that time beyond question. If this sounds counterintuitive,
it is only because in the centuries between then and now a dramatic inversion of the relative status of science and of religion has taken place, so
that it is far more likely now for religion to be seeking to secure its legitimacy by appealing to science than the reverse. Indeed, for some, religion
is legitimate only to the extent that it can demonstrate its credentials on
the terms set out by science. The situation in the seventeenth and eighteenth centuries was rather different.

In focusing on the question of the utility of the new sciences, Gimcrack's remark went to the heart of one of its major weaknesses: it was
regarded by many as useless. If the mockery of its detractors offers one
measure of its vulnerability on this point, the defensive reaction of proponents of the new philosophizing offers another. Apologists for the new
science, in England at least, became increasingly concerned to establish
its usefulness, and the latter half of the seventeenth century is witness to
numerous defenses of experimental natural philosophy on the basis of its
practical utility. In 1663, Robert Boyle wrote an entire work devoted to the
usefulness of experimental natural philosophy. Thomas Sprat's *History of
the Royal Society* (1667) is written in the same vein. While the title might
suggest to modern readers a disinterested chronicle of the early years of
this fledgling scientific organization, the book was, in fact, a highly par-
tisan account. The entire third section, over one hundred pages, is given
over to an enumeration of the practical benefits of the society's initiatives.
Another divine and fellow of the society, Joseph Glanvill (1636–80), wrote
no fewer than three works in the 1660s and 1670s seeking to establish the
manifold ways in which the new philosophy had contributed to the welfare of society.[27] These defensive publications indicate that criticisms of

the utility of science were taken seriously enough to warrant robust responses.

These arguments for the usefulness of natural philosophy attempt to address both old and new conceptions of utility, and do so in three ways. Joseph Glanvill pointed out that "the *Real Philosophy* . . . will assist and promote our *Vertue*, and our *Happiness;* and incline us to imploy our selves in living according to it."[28] By "real" philosophy he means the new experimental philosophy that deals with "real" things rather than notions or definitions. The new way of philosophizing thus encompasses the moral aims of traditional philosophy. Beyond this, experimental natural philosophy is religiously useful because it establishes premises for natural theology:

> the study of Nature and God's works, is very serviceable to Religion. . . . and
> the *Divine Glory* is written upon the *Creatures*, the more we study them, the
> better we understand *those* Characters, the better we *read* his *Glory,* and
> the more fit we are to celebrate, and proclaim it. Thus the *knowledge* of
> *God's Works* promotes the *end of Religion.*[29]

This second argument, for religious usefulness, was crucial in establishing the religious legitimacy of the new science, which, as the attacks of Swift and others illustrate, was then in a vulnerable position. But in a third argument Glanvill insisted on a new understanding of usefulness, one that is to do with what Bacon designated "the relief of the human estate": "I say then, That it was observed by the excellent Lord *Bacon,* and some other *ingenious Moderns,* That *Philosophy,* which should be an *Instrument* to *work* with, to find out those *Aids* that *providence* hath laid up in Nature to help us against the *Inconveniences* of *this* State, and to make such applications of things as may tend to a *universal benefit.*"[30]

If we ask ourselves what has changed in past three hundred years in terms of our perception of the usefulness of scientific knowledge, it is not primarily that our science yields practical benefits in a way that early modern science did not (although in some sense this is true). Rather, a profound change has taken place in what counts as useful knowledge—a change in which the moral and religious elements to which Glanvill alludes have simply dropped out of the frame. In the late seventeenth and early eighteenth centuries, however, there were ongoing debates about the moral role of natural philosophy, and the initial attempts of Royal Society apologists to vindicate experimental philosophy elicited considered counterarguments that went well beyond the casual mockery of its more popular detractors. Two lines of argument were articulated. First, it was claimed that an emphasis on the material benefits of learning runs counter to the whole point of accumulating knowledge, which is the moral shaping of the individual. In other words, the older criterion of usefulness, understood as the nurture of moral and religious qualities, was reasserted. Second was the charge that even on its own flawed conception of usefulness, experimental philosophy had failed, since it had fallen lamentably short of offering any significant practical innovations that genuinely promoted human welfare.

The clearest articulation of the arguments from the other side came from the pen of Meric Casaubon (1599–1671), son of the famous classicist Isaac Casaubon, and a distinguished scholar in his own right. Responding to Glanvill's claims for the usefulness of the new philosophy, Casaubon posed the key question: What is meant in these discussions by "useful" and "useless"?[31] The tending of furnaces and raking over the entrails of men and beasts, he contended, will never make men wise, "nor ever prove of any great use."[32] Considering the subsequent development of science and technology, we may be tempted to think that Casaubon was lacking in prescience. But his point was that whatever practical applications to human welfare might result from experimental science, these would not establish its usefulness in the true sense of the word. For, as he puts it: "To *moralize* men is the best *use* of any worldly thing which can be made."[33] A similar criticism was voiced by Henry Stubbe (1631–76), physician, controversialist, and second keeper of the Bodleian Library. (Among his diverse accomplishments was authorship of the first treatise on chocolate.) Stubbe argued that the pursuit of experimental science was incapable of producing "that Moral discipline which instructs us in the nature of virtue and vice, of Distributive and Commutative Justice: humane actings and the due course, as well as exorbitances of our passions." These should be the focus of philosophical attention rather than "aphorisms of Cider, planting of Orchards, making of Optick Glasses, magnetic and hortulane Curiosities" (see figure 7).[34] Both Casaubon and Stubbe deferred to Aristotle in their defense of the moral goals of philosophy. Later Jonathon Swift would echo their concerns, contending that the much-vaunted indicators of material progress—the compass, gunpowder, and printing—were the products of an age of ignorance.

Not surprisingly, these reassertions of more traditional goals of learning had implications for the new understanding of human progress. For Casaubon an idea of progress that relied on cumulative human endeavors was essentially Pelagian in its import, for it arrogates to itself the role that rightly belongs to divine providence:

> God knows his own time, and when that time is come, one single man
> shall do more perchance, then a combination of many thousands, at
> another time. [A man might give a profitable reason, now it is done,
> why God would reserve the discovery of the New world lately found out,
> to these times. . . . The invention of Printing, did much promote learn-
> ing (good useful learning) that reformation which God intended in his
> Church.] Every thing hath its proper time, and that time is a secret of
> God's dispensation.[35]

This is an explicit rejection of the new idea of progress championed by Bacon and the Baconians of the Royal Society, according to which progress consists in the incremental augmentation of a body of knowledge by successive generations of experimental philosophers.

It might be argued that all the defenders of the new science were doing was reiterating standard defenses of the moral and practical utility of natural philosophy. But now these two elements of usefulness were separable, for the practical outcomes were arrived at collectively, and hence were separable from the moral probity of any single individual involved in the process. Previously, in order to be the recipient of veridical knowledge a certain work had first to be performed on the self. Now, the philosophical regimen that had once been directed toward the moral shaping of the person was objectified into an experimental program that was indifferent to the moral character of those who pursued it. The hierarchy of

SYLVA,

Or A DISCOURSE OF

FOREST-TREES,

AND THE

Propagation of Timber in His
MAJESTIES Dominions.

As it was Deliver'd in the *ROYAL SOCIETY* the xv[th] of *October*, CIɔIɔCLXII. upon occasion of certain *Quæries* propounded to that *Illustrious Assembly*, by the *Honourable* the Principal *Officers*, and *Commissioners* of the Navy.

To which is annexed

POMONA; Or, An *Appendix* concerning *Fruit-Trees* in relation to *CIDER*; The *Making*, and severall wayes of *Ordering* it:

Published by expresse Order of the ROYAL SOCIETY.

ALSO

KALENDARIUM HORTENSE; Or, the *Gard'ners Almanac*; Directing what he is to do *Monthly* throughout the *Year*.

All which severall *Treatises* are in this *SECOND EDITION* much *Inlarged* and *Improved*

BY

JOHN EVELYN *Esq*; Fellow of the *ROYAL SOCIETY*.

——*Tibi res antiqua laudis & artis*
Ingredior, tantos ausus recludere fonteis. Virg.

LONDON,
Printed for *Jo. Martyn,* and *Ja. Allestry,* Printers to the *Royal Society.* MDCLXX.

Figure 7. John Evelyn, *Sylva* (2nd ed. 1670). This edition includes a Gardener's Almanac and an appendix on cider making. These kinds of activities, associated with the Royal Society, attracted the disdain of those who believed that philosophers should be engaged in more edifying endeavors.

uses of natural philosophy—of the kind that we saw in Simplicius—also collapses, as does the idea of natural philosophy as a preparation for the study of the higher sciences. This allowed for a future justification of natural philosophy, or "science" as it would eventually become, on the basis of its practical utility alone.

The Relief of the Human Estate

A third illustration of the general trend toward objectification of the virtues, and one linked to the new notion of progress, is the case of Christian charity. In the famous thirteenth chapter of Saint Paul's first letter to the Corinthians, charity (or love, Gk. *agapē*) is identified as the greatest in the triad of virtues: faith, hope, and charity. In Thomas Aquinas these three are formally identified as "theological virtues," and in the Middle Ages these specifically Christian virtues supplemented the list of traditional moral and intellectual virtues enumerated by Aristotle.[36] Of the theological virtues, as we have already seen, "faith" is transformed from something like personal "trust" into what will eventually become a kind of epistemic vice: faith, in the modern period, entails belief in certain propositions. Understood pejoratively, faith is belief in the absence of requisite evidence. Charity also undergoes a major reconfiguration. In much the same way that changes in the meaning of "religion" are signaled by the addition of an article, we can see in the earliest uses of the idea of "*a* charity" the objectification and institutionalization of what had once simply been an inherent quality.

One of the very first references in the English language to "a charity," understood as an institution rather than a virtue, appears in John Evelyn's memoirs. Evelyn was an active member of the early Royal Society and a dedicated diarist. In an entry dated 10 March 1687, Evelyn recounts visiting Christ's Hospital, which he refers to as a "noble, pious, and admirable charity."[37] The history of this "charity" is itself quite telling. Originally founded in the sixteenth century, Christ's Hospital had been dedicated to the education of the poor, utilizing property that had become available following the dissolution of the monasteries. The school was thus, quite literally, the product of the process of secularization, understood in its original

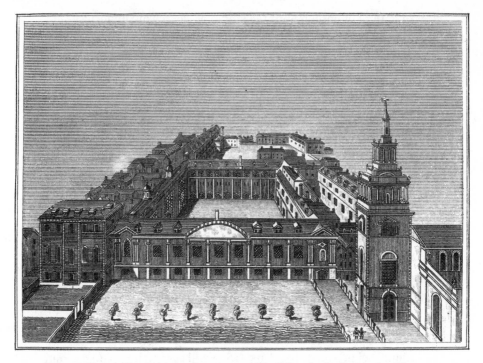

Figure 8. Christ's Hospital in London, 1770, with buildings by Sir Christopher Wren and Nicholas Hawksmoor. The charity school is now located near Horsham in West Sussex.

sense as the conversion of ecclesiastical property to secular use. The new institution also exemplified the disciplinary society, since it extended an original model of monastic discipline to a particular segment of society, and did so with the goal of improving social welfare. (Evelyn had remarked on its "order, economy and excellent government.") The particular form of discipline imposed in this charitable institution came to be connected with the new sciences. When Christ's was refounded by Charles II in 1673, it became the site of the Royal Mathematical School, which trained mathematicians and navigators. From the very beginning of its second foundation the institution had a close relationship with the Royal Society. Samuel Pepys, diarist, naval administrator, and sometime president of the Royal Society, was for a time its vice president, and distinguished scientists such as Isaac Newton, John Flamsteed, and Edmund Halley made contributions to its curriculum.[38] The designation of this institution as a charity is emblematic not only of important changes in understandings of charity and its relation to the idea of usefulness, but also in how charity was brought into a relation with the new sciences.

Charity was already a central theme in Francis Bacon's case for a new kind of philosophy.[39] In the concluding remarks of his Preface to *The Great Instauration* (1620) Bacon inquires after "the true ends of knowledge," maintaining that those who pursue knowledge must do so not "for pleasure of the mind . . . but for the benefit and use of life; and that they perfect and govern it in charity."[40] Bacon's own rendition of the double injunction to love God and love neighbor, refracted through his advocacy of the new philosophy, is this: "the glory of the creator and the relief of man's estate."[41] Elsewhere "love of neighbour" is alternatively rendered by a newly coined English word—"philanthropy." Bacon describes philanthropy as the greatest "of all the virtues and dignities of the mind" and which is nothing less than "the character of the Deity"; it is a form of goodness that equates to "the theological virtue Charity."[42] Insofar as the new science is directed toward the relief of man's estate—and this, for Bacon, is its true end—it becomes the very embodiment of charitable activity:

> But yet evermore it must be remembered that the least part of the knowledge passed to man by this so large a charter from God must be subject to that use for which God hath granted it; which is the benefit and relief of the state and society of man; for other wise all manner of knowledge becometh malign and serpentine, and therefore as carrying the quality of the serpent's sting and malice it maketh the mind of man to swell; as the Scripture saith excellently, *knowledge bloweth up, but charity buildeth up.*
> (I Cor. 8:1)[43]

Bacon thus resists one obvious reading of this maxim from I Corinthians, in which knowledge is unfavorably compared with charity, claiming instead that scientific knowledge is exempt from this biblical censure precisely because of its potential to ameliorate the material infirmities that attend the human condition. But what he also attempts to do here is to align the traditional philosophical quest, one that focused on the moral formation of the individual, with "puffed up knowledge." In one sense he endorses the Augustinian priority; in another, he represents a significant departure from it (or at least gives legitimacy to a certain line of departure from it). In these sentiments there is also an incipient Pelagianism—the doctrine, roundly opposed by Augustine and decried by Protestant theologians, according to which human beings can work their way out of their sinful condition through their own efforts. As we have seen, Bacon regards human infirmity as resulting from the Fall, and the need to relieve the human estate is a direct consequence of the primeval defection of Adam. Science is depicted as offering a means of human redemption alongside religion, which is assigned the complementary role of redressing the moral corruption that followed the Fall. Bacon's successors in the Royal Society also made the connection between science and charity in their defenses of the usefulness of their activities. Glanvill wrote in an essay on "The Usefulness of Real Philosophy to Religion" (1676) that the experimental study of nature "promotes the *end of Religion*" and the life of virtue.[44] This virtue is in turn accompanied by acts of charity, giving rise to "*Invention of Arts*, and *Helps* for the *benefit* of *Mankind*."[45] Other defenders of the moral and practical usefulness of experimental natural philosophy mounted similar arguments.[46]

These reconfigurations of the virtue of charity and its new associations with utility did not go unnoticed or unchallenged. In his censure of the priorities of the early Royal Society, Henry Stubbe maintained that the linking of charity with the new science directly contradicted Saint Paul's seminal description of Christian charity: "Though I have the gift of Prophecy, and understand all mysteries, and all knowledge . . . and have not charity, I am nothing" (I Cor. 13:2-3).[47] Coupled with Pauline doubts about "the wisdom of the world," passages such as these suggested that scientific knowledge was in a completely different category from Christian charity. But as we have seen, practitioners of the experimental philosophy could respond that scientific pursuits actually satisfied another biblical injunction to do with charity—the command to love God and love neighbor. Natural philosophy fulfilled the former, because it uncovered evidence of divine wisdom in the frame of nature, thus promoting love of

God; the latter, because the material contributions of science were to be understood as manifestations of love of neighbor.

The fortunes of charity in this period are consistent with a general trend to redefine the virtues, or to reduce them to duties or requirements of moral law. In the case of charity, the focus shifts from the moral condition of the individual to the performance of duties toward others. For Aquinas, charity had been understood primarily in terms of friendship with God, for the sake of which it was extended to our neighbors.[48] In the early modern period, it is increasingly duties to neighbors that become the primary concern. This is evident not only in Bacon, but also in the treatments of such prominent natural law theorists as Hugo Grotius (1583–1645) and Samuel Pufendorf (1632–94), who deal directly with moral philosophy and in whom also we witness the beginnings of the secularization of charity.[49] Pufendorf understands charity as encompassing the duties of compassion, liberality, beneficence, gratitude, and hospitality. Charity is an "imperfect duty" that is not absolutely necessary for the preservation of mankind, although it "embellishes" human society and makes it more commodious.[50] The "law of humanity or charity" then, supplements the more binding considerations of justice that operate quite independently of the moral status of those enacting them.[51]

The responses to questions about the usefulness of scientific knowledge considered above also point to the beginnings of a depersonalized, objective understanding of the uses of *scientia*. Admittedly, there remains a somewhat precarious balance between the formative and utilitarian aspects of knowledge. This is reinforced with appeals to familiar religious dichotomies—faith and works, love of God and love of neighbor, edification of the soul and preservation and maintenance of the body—with science being identified with good works, love of neighbor, and promotion of the health of the body. But these defenses set the stage for the subsequent displacement of the character-related aspects of the activity by an new emphasis on its practical outcomes. This is partly owing to the fact that the arguments for the moral usefulness of science—that is, the claim that experimental knowledge of nature will "assist and promote our virtue" as Joseph Glanvill put it—were themselves premised on a conception of the moral order of the cosmos that ultimately proved to be incompatible with the new science. For a start, the new sciences entailed an explicit rejection of the symbolic significance of nature and, to some extent, of its tropological character as well. Echoes of the traditional argument of the kind we heard in Ptolemy, for whom the study of the cosmos reforms human nature and reduces it to a spiritual state, now sound a little hollow. The new Copernican world was far removed from the cozy, medieval, Ptolemaic world in which moral meanings were inscribed on the cosmos. Pascal seemed acutely conscious of this, famously remarking that the eternal silence of the infinite heavens filled him with dread.[52] The vast expansion of the cosmos, the idea of empty space, the evacuation from nature of transcendental meanings, all presaged a mute universe that seemed to offer no moral guidance to those who studied it. Pascal's considered remark was an anticipation of a common present-day scientific sentiment, memorably encapsulated by Steven Weinberg in these words: "The more the universe seems comprehensible, the more it also seems pointless."[53] From this period onward, the study of nature is pursued not only for the purpose of moral edification, but also to yield technologies that will offer ways of relieving the human condition. In time, the latter will completely displace the former.

Self-Dominion and Dominion over Nature

There is yet one more conception, briefly mentioned in the previous chapter, that I wish to explore in setting out this new impersonal conception of progress. It, too, is a familiar Judeo-Christian conception—that of dominion over nature. At this time, it is also reconfigured, in the familiar pattern, from being primarily an internal matter to one that is redirected outward toward the natural world. The biblical imperative for human beings to exercise dominion over nature has enjoyed considerable notoriety over the past fifty years, largely on account of the perception that it is deeply implicated in environmental degradation. The historian Lynn White Jr. did more than anyone to make this connection, suggesting in an influential paper in the 1960s that on account of the biblical injunction to exercise dominion over nature, the Judeo-Christian tradition "bears a huge burden of guilt for environmental deterioration."[54] Although the White thesis has been subjected to searching criticism, to a degree it is on the right track. The one key element missing from White's analysis is the fact that this biblical injunction also underpinned the modern scientific project, providing an important source of motivation for the investigation of nature and giving religious legitimacy to a project that, as we have seen, was more vulnerable in its early stages than we have sometimes assumed. The other weakness of the White thesis, and the one that most concerns us here, is its incomplete analysis of the influence and meanings of the biblical motif of dominion before the rise of science.[55]

As noted earlier, patristic and medieval exegetes frequently interpreted the imperative to exercise dominion over the beasts as an advocacy of psychological mastery over bestial passions. The exercise of dominion was thus understood figuratively, and the human person was regarded as a microcosm. Origen thus wrote: "Understand that you have within yourself herds of cattle . . . flocks of sheep and flocks of goats . . . and that the birds of the air are also within you. . . . you are another little world."[56] From this perspective, the injunction to exercise dominion was understood as a command to exercise of self-control, with the animals representing, variously, "the inclination of the soul and the thought of the heart," or "bodily desires and the impulses of the flesh."[57] This was a standard reading of the meaning of dominion. Augustine, for example, described the sanctified soul in these terms: "Then the wild animals are quiet and the beasts are tamed and the serpents rendered harmless: in allegory they signify the affections of the soul." He continues: "So in the 'living soul' there will be beasts that have become good by the gentleness of their behaviour. . . . For these animals serve reason when they are restrained from their deathly ways."[58] To a degree, these patristic notions paralleled the Platonic idea that the whole cosmos, including the irrational creatures, was teleologically ordered to promote the progress of the human soul.

The idea of a mastery of external things was not entirely absent from earlier discussions of the idea of dominion. However, Christian authors agreed that following the Fall, Adam had at the same time lost his original capacity to control his passions, and his exterior dominion over the beasts. The goal of the religious life was to restore this lost dominion, but the priority was always the reestablishment of the proper hierarchical relations in the soul. This inner restoration was a necessary precondition for the reassertion of human dominion over things. As Aquinas put it: "Man in a certain sense contains all things; and so according as he is master of what is within himself, in the same way he can have mastership over other things."[59]

A new set of priorities comes into play in the early modern period. For a start, the idea that the human person was a microcosm fell into decline. At the same time, as we have seen, a change of emphasis in the sphere of biblical hermeneutics saw a renewed focus on a literal reading of the Genesis narratives. This meant that the fanciful allegorizing upon which rested the idea of the beasts as representations of states of the human soul began to fall from favor. As a further consequence, dominion over nature was increasingly understood, quite literally, as the exercise of control over the natural world. Once again, Francis Bacon offers us an instructive example. The injunction "have dominion" was to be satisfied by the accumulation of scientific knowledge: "by digging further and further into the mine of natural knowledge" human investigators could extend "the narrow limits of man's dominion over the universe" to their "promised bounds."[60] In the closing lines of the *Novum Organum* Bacon sets this whole program in its religious context:

> For man by the fall fell at the same time from this state of innocency
> and from his dominion over creation. Both of these losses however can
> even in this life be in some part repaired; the former by religion and faith,
> the latter by arts and sciences. For creation was not by the curse made
> altogether and forever a rebel, but . . . is now by various labours . . . at
> length and in some measure subdued to the supplying of man with bread;
> that is to the uses of human life.[61]

Notice that whereas Aquinas had suggested that the moral and religious task of self-mastery would provide the means of a reestablished mastery over nature, Bacon keeps these two things distinct. Religion and faith offer the means of redressing the psychological damage wrought by the Fall and of restoring our self-mastery, but it is the arts and sciences that are to be pursued if we are to reattain our literal mastery of nature.

The theme of a restored dominion over nature became a commonplace in early modern justifications of the new science. Thomas Sprat wrote in his *History of the Royal Society* that one of the objectives of the fellowship was the reestablishment of "Dominion over things."[62] Joseph Ganvill agreed that the new philosophy would afford "ways of *captivating Nature, and making her subserve our purposes and designments.*" This, he confidently asserted, would lead to the restoration of "the *Empire of Man over Nature.*" For Glanvill, it was a religious imperative that nature be "master'd, managed, and used in the Services of Humane life." Such services might include "the accelerating and *bettering of Fruits, emptying Mines, drayning Fens* and *Marshes.*" "*Lands,*" he concluded, "may be *advanced* to scarce credible degrees of *improvement,* and innumerable other *advantages* may be obtained by an *industry* directed by *Philosophy* and *Mechanicks.*"[63] Philosophy was never so useful.

Again, this redirected dominion is consistent with the new understanding of the goals of philosophy. Bacon contended that the true end of philosophy, and more specifically natural philosophy, is the transformation of the world, rather than the soul of the philosopher. This requires an outwardly directed dominion. He writes: "Knowledge is to be sought not for the quiet of resolution but for a restitution and reinvesting (in great part) of man to the sovereignty and power . . . which he had in his first state of creation."[64] Here Bacon has in mind the restoration of Adam's original encyclopedic knowledge of nature and, equally, his dominion over it. Bacon thus rejects the idea that the philosopher is to be concerned first and foremost with the quiet consolations of the contemplative life. Rather, the philosopher is to be engaged in a mission to restore the natural world to its original perfection, or at least something approaching it. This is a radically new conception of the goals of natural philosophy (as Macaulay pointed out). Stephen Gaukroger has more recently argued that Bacon represents "the first systematic comprehensive attempt to transform the early modern philosopher from someone whose primary concern is with how to live morally into someone whose primary concern is with the understanding of and reshaping of natural processes."[65] Self-dominion, as the goal of the natural philosophical life, is eventually displaced by the quest for a dominion over things.

As a reminder of the power of the classical conception of philosophy, and how it contrasts with this new orientation, we need look no further than Seneca's observation that his own time had seen the invention of transparent windows, central heating (of a kind), and shorthand. But, he declares, "the inventing of such things is drudgery for the lowest slaves; philosophy lies deeper. It is not her office to teach men how to use their hands. The object of her lessons is to inform the soul."[66] Bacon's repudiation of this understanding of the goals of philosophy was taken up by a number of seventeenth-century natural philosophers. Thomas Hobbes writes in the *Leviathan* (1651) that "the felicity of this life consisteth not in the repose of a mind satisfied."[67] Philosophy, he says elsewhere, is not about "the inward glory and triumph of mind that a man may have for the mastering of some difficult and doubtful matter." Rather the goal of philosophy is "that we may make use to our benefit of effects formerly seen . . . for the commodity of human life."[68] René Descartes also speaks of a new practical philosophy "which would be very useful in life." Through this philosophy, says Descartes, we will "make ourselves, as it were, the lords and masters of nature." Finally, Robert Boyle noted that "the barren philosophy, wont to be taught in the schools, hath hitherto been found of very little use in human life." The purpose of philosophy was now to produce useful knowledge, to master and command nature, and to extend the human empire over it.[69] Each of these thinkers, in his own way, makes the point that natural philosophy is not about cultivating the virtue of *scientia*, but is about the practice of an activity that in many ways resembles modern science.

In sum, the motif of "dominion over nature" undergoes a similar mutation to the ideas of progress, utility, and charity that we have considered in this chapter. Again we witness a change of emphasis away from the interior psychological aspects toward the external world. This shift is directly linked to the modern idea of progress, which is now very much concerned with the restoration of the empire of man over nature, and the ongoing expansion of those imperial boundaries. In time, of course, the rhetoric of "restoration," which drew its potency from Christian ideas of creation, Fall, and redemption, faded from the picture. The idea of a partial return to a paradisal perfection was replaced by a new idea of progress that had originally been justified in religious terms, but increasingly assumed a normative value of its own. Progress had once been understood as progress toward a particular end. Now it had become an end in itself.[70]

:::

The themes of this chapter, like those of the previous one, point in their own way to the consequences of the demise of a long-standing Aristotelian synthesis of metaphysics and ethics. With the decline in virtue ethics, the modern alternatives of deontology (which understands ethical obligation in terms of duties) and utilitarianism (which focuses on the outcomes of acts) begin to come to the fore. We see this most clearly, perhaps, in the case of charity. Once the preeminent theological virtue, charity is no longer an infused *habitus* that makes the individual more Christlike. Now it is understood in terms of duties one has to others (deontology), or in terms of outcomes that promote human welfare in some measureable material way (utilitarianism). These transformations make possible "charity" in the modern sense—the sum total of a number of beneficent acts that promote material welfare. The virtue of charity, insofar as it persists at all, is distributed across particular organizations—"charities"—and those who contribute to them. Like *scientia* and *religio, caritas* begins to lose its anchorage in the human heart and become objectively identifiable with actions or institutions.

It is also worth pointing out that the new way of understanding progress, when applied to a reified "religion," necessarily places it at something of a disadvantage to a reified "science." When *scientia* and *religio* were both virtues, progress in each case amounted to much the same thing—namely, a kind of training of the mind or soul that moved the person in the direction of their natural fulfillment. When science and religion become reified in a way that privileges doctrinal content, however, progress will mean something quite different in each case. Viewed simply, the doctrines of Christian religion cannot be indefinitely augmented in the way that the doctrines of science can, since there is a degree of finality and sufficiency in the original deposit of revealed religious truths. Admittedly, a form of this difficulty had already been acknowledged by the Fathers and the scholastics, and efforts had been made to argue for the progressive nature of religion. But even the most progressivist understanding, promoted by the fifth-century monk Vincent of Lérins, had spoken of "the same teaching, in the same sense, with the same meaning."[71]

From the eighteenth century onward the idea of progress is one that tends to be associated with science. Arguably, this association was first made by Protestant interpreters of history, who linked the reformation of religion to a subsequent reformation of the sciences, and who regarded both developments as part of a providentially inaugurated phase of history. The Puritan divine Cotton Mather thus wrote in 1701 of the "incredible darkness" that had fallen over western Europe following the fall of Rome, and of the fact that "*Learning* was wholly swallowed up in *Barbarity*." This was brought to an end "by the revival of letters . . . which prepared the whole world for the *Reformation* of *Religion* too, and for the *Advances of the Sciences* ever since." Numerous Protestant commentators, including Francis Bacon and Thomas Sprat, had made similar connections between the reformation of religion and the progress of learning.[72] These "Protestant" interpretations of history, which attributed the impediment of the sciences to the Catholic Church, offered a ready-made narrative to Enlightenment critics of Catholicism who often turned out to be less discriminating in their attribution of blame. Nicolas de Condorcet, whose *Sketch for a Historical Picture of the Progress of the Human Spirit* (1795) was a paradigmatic statement of the Enlightenment ideal of progress, claimed that "the triumph of Christianity had been the signal for the complete decadence of philosophy and the sciences."[73] To be sure, it was *Catholic* Christianity that played this role in the Middle Ages. But Catholicism could easily become the more generic "religion." Thus, when history is reconstructed on the basis of an assumption of the inevitability of progress, the question of science's origins and development is typically framed not in terms of the unique and contingent conditions that make the emergence and persistence of science possible. Rather the question is posed in this way: given the intrinsically progressive nature of science, what are the factors that have inhibited its natural flourishing? The answer for many came in the form of "religion," understood as a historical reality in which the differences between rival versions—Protestant or Catholic—were less significant than what they shared: namely, an inflexible commitment to a set of unchanging truths derived from authorities. In this version of things, not merely is science itself progressive, but its appearance on the stage of history signals a new and more advanced phase of social development. This view of history was powerfully captured by Comte, who saw religion as a primitive stage through which humanity must pass to achieve scientific maturity. Few, if any, historians now subscribe to this crude historicism, but this pattern of progress nonetheless informs many present-day assumptions about the future of science and religion.

Finally on the question of progress, it is worth pointing to the fact that progress is an axiological concept—that is to say, it draws upon the realm of values. In order to see the force of this, consider Macaulay's account of Bacon's achievement, set out at the beginning of the chapter. On Macaulay's version of events, natural philosophy had disdained to concern itself with human welfare. Then along came Bacon with the new idea that natural philosophy should concern itself with "the good of mankind." Why, we might reasonably ask, had no one thought of this before? The fact is that considerable thought had been given to this question, it is just that premodern thinkers had subscribed to a different notion of the where human good lay, and they had pursued that. Bacon's significance, considered in this light, is to have appealed successfully to a different set of values, albeit values there were incipient in the tradition. His strategy, following the reform of religion in the previous century, was to argue that a genuinely Christian conception of human good required a privileging of charity—understood in practical terms—over wisdom. (While such a priority had already been argued for in the Middle Ages, the key difference lay in the new understanding of charity.) This stance was underscored with a view of sacred history that saw in natural philosophy the means both to restore the earth to its Edenic state and to prepare humanity for the impending end of history.[74] The claim that no one before Bacon had thought of using natural philosophy to "benefit mankind" is thus mistaken. The real issue was to do with alternative conceptions of the ultimate human good—the progress of the human soul or the material betterment of mankind—and it is worth reminding ourselves that it is not a straightforward matter to adjudicate between them.

It also follows from this that there is something inherently unstable in the modern understanding of progress. Progress had once been thought of as the movement of human beings toward certain given ends. But without at least an implicit teleology (which was precisely what the new natural philosophical approaches sought to dispense with) the notion of progress is difficult to sustain. Progress, in other words, is goal dependent; progress is *toward* some end. Without goals, progress is just change. This was less of a problem for those whom we regard as early modern pioneers of progress. As we have just observed, individuals such as Bacon understood themselves as participants in a providential plan, and indeed often saw their mission more in terms of reform than progress. In this sense, their utopian goals were consistent with their underlying values. The vestiges of these values have carried over into the present scientific enterprise, but arguably without any of the assumptions that would make a belief in progress rational.[75]

[6]

Professing Science

When one turns to the magnificent edifice of the physical sciences,
and sees how it was reared; what thousands of disinterested moral
lives of men lie buried in its mere foundations; what patience
and postponement, what choking down of preference, what
submission to the icy laws of outer fact are wrought into its very
stones and mortar; how absolutely impersonal it stands in its vast
augustness.
—Williams James, *The Will to Believe*

"First accumulate a mass of Facts: and then construct a Theory."
That, I believe, is the true Scientific Method.
—Lewis Carroll, *Sylvie and Bruno*

Oceania was at war with Eastasia: Oceania had always been at war
with Eastasia.
—George Orwell, *1984*

In an 1867 article written for the *Dublin Review*, William George Ward
offered the following definition of science: "We shall, for convenience'
sake, use the word 'science' in the sense which Englishmen so commonly
give to it; as expressing physical and experimental science, to the exclu-

sion of theological and metaphysical."[1] Ward was a Catholic convert and religious writer, best known today as the sometime editor of the *Dublin Review* and cofounder of the Metaphysical Society, whose membership was a virtual roll call of the leading intellects of Victorian England. His article, it must be said, was largely unremarkable save for the fact it provided one of the first references in the English language to a newly emerging and restricted sense of the term "science."[2] Ward suggested that "science" was now often used by the English to refer only to the physical and experimental sciences, and that those activities were themselves practiced in ways that made no reference to metaphysical or theological concerns. Such a stipulation was necessary because, although this usage was now starting to become common, "science" was still then understood in a wide variety of senses.

As we saw in chapter 4, the disappearance of the idea of *religio* as an interior virtue was accompanied by the emergence of our modern notion "religion." But the parallel demise of the intellectual virtue *scientia* had a different kind of outcome for it did not immediately result in the appearance of our modern idea "science." Instead the term "science" continued to refer to knowledge in general, and the formal properties that Aristotle had attributed to scientific knowledge—logical demonstrability and certainty—were mostly regarded as irrelevant to the new scientific practices that focused on experiments and the accumulation of observational data. In the case of *religio*, the label for the virtue had been almost seamlessly reapplied to the relevant body of knowledge. In the case of *scientia*, however, the label was at first not redeployed to describe those activities that we would now regard as "science." Instead, the main ways of designating the study of nature continued to be "natural philosophy" and "natural history," and although these activities were connected to various moral concerns, they were no longer linked to some supposed virtue *scientia*. Moreover, and somewhat curiously from our present perspective, neither were these practices regarded as "scientific" in the formal, traditional sense.

However, this situation was not to continue indefinitely. In the nineteenth century, "science," as we now understand it, came to be constructed in a way that resembled the early modern construction of the idea "religion." That is to say, it was aggregated from a range of activities and distanced from the personal qualities of those who practiced it. This took place when "science" came to be linked to a putatively unified set of practices ("the scientific method"), associated with a distinct group of individuals ("scientists"), and purged of elements that had once been regarded as integral to its status and operations (the theological and the metaphysical). Modern religion had its birth in the seventeenth century; modern science in the nineteenth. Properly speaking, then, this belated appearance of "science" provides the first occasion for a relationship between science and religion.[3]

The Meanings of Science

For much of the nineteenth century "science" was used in a wide variety of ways. Only four years before Ward offered his stipulation definition, the news correspondent Edward B. Freeland grumbled that the term "has been so indiscriminately applied to very diverse departments of our intellectual domain, that it has ceased to have any distinctive or well-defined signification."[4] Even a cursory survey of definitions and usages of the word bears out his complaint. "Science" was still used to refer simply to systematic knowledge in general. In 1828, for example, the most widely read literary magazine of the period, the *Athenaeum*, divided the sciences into "exact, experimental, speculative, and moral." William Whewell—polymath, master of Trinity College, fellow of the Royal Society, and founding fellow of the British Association for the Advancement of Science—classified natural theology among the inductive sciences, and there were still those in the first decades of the nineteenth century for whom theology retained its status as "the first and greatest of the sciences."[5] In a more technical sense, "science" might refer to a process of logical demonstration as the relevant entry in the first edition of the *Encyclopaedia Britannica* (1771) testifies: "SCIENCE, in philosophy, denotes any doctrine, deduced from self-evident and certain principles, by a regular demonstration."[6] This definition bears the obvious trace of the older ideal of science that dates back to Aristotle's *Posterior Analytics*, but it was clearly not applicable to what we would now regard as sciences.

There were also more narrow and even idiosyncratic understandings of the word. In nineteenth-century Oxford, "science" could refer to ethics or the study of Greek philosophy. E. A. Freeman, in his recollections of mid-century Oxford was to observe of an acquaintance: "I remember him years ago as a logic and science coach. I don't mean for cutting up cats, but what science meant then, Ethics, Butler, and such like."[7] The term had an even more specific reference to the study of the classics. "Science," said Mark Pattison, reflecting on his period as rector of Lincoln College, had meant "attainment in Aristotle."[8] A contributor to the *Athenaeum* concurred that at New College, Oxford, in the middle decades of the nineteenth century, Greek philosophy was referred to as "science."[9] Further, as we shall see, at this time the profession of "scientist" did not yet exist, and neither was there a strong representation of what we would call the natural sciences in the university curriculum—at least not at the ancient universities of England where classics, law, and theology held sway. At Oxford in 1870 there were 145 classics fellowships as against four in the natural sciences.[10] The preference for the "morally useful" studies, evident in seventeenth-century reactions against the new experimental science, thus continued to prevail to a significant degree. In short, there was neither a vocational identity nor specific institutional contexts that might lend "science" a more concrete identity.

A Common Context?

The idea of a study of nature that excluded the metaphysical and theological (to invoke Ward's definition of science) was profoundly at odds with a common understanding of natural philosophy and natural history that had prevailed up until the mid-nineteenth century. As already noted in chapter 4, no less a figure than Isaac Newton, in a famous passage in the second edition of the *Principia,* had announced that discourse about God is a genuine part of natural philosophy. This assertion alone should be sufficient to give pause to those who too readily assume a close affinity between the "science" of Newton and science as we presently conceptualize it. To be sure, there remain important questions about the identity of early modern natural philosophy and the extent to which it was theologically oriented.[11]

But the general point remains that the Newtonian stance was not uncommon. Furthermore, this religious orientation was not simply a matter of adding pious glosses to an independent body of knowledge. From the beginning of the seventeenth century to the middle of the nineteenth, natural theology provided a vital unifying theme for natural history and natural philosophy. Noël-Antoine Pluche, whose eight-volume *Spectacle de la nature* was one of the most widely translated works of eighteenth-century natural history, spoke for the majority of early modern naturalists when he asserted that "all nature is link'd together by one universal Law of Harmony and Agreement . . . as the Whole Earth declares itself to be the Work of one only all-wise Creator."[12] The centrality of the idea of contrivance or design, along with the concept of divinely imposed and universal laws of nature, was central to much of the "scientific" enterprise up until the middle decades of the nineteenth century. Without this overarching theological framework, the sciences of nature ran the risk of becoming mere catalogs of undigested and unrelated facts.

One concrete indication of the unifying power of natural theology comes from the fact that at Cambridge, from the late eighteenth century onward, the works of William Paley formed the cornerstone of the curriculum.[13] These works included *The Principles of Moral and Political Philosophy* (1785) and *The Evidences of Christianity* (1794). The first of these set out a theory of utilitarian ethics similar in some respects to better-known system of Jeremy Bentham, although the two men had arrived at their conclusions independently. *The Evidences* was compiled from a range of standard eighteenth-century arguments dealing with miracles and prophecy, and presented what was essentially a digest of external evidences for Christianity. To these two works was later added the equally influential *Natural Theology* (1802), which contained the famous design argument based on the analogy between a watch and the natural world. The latter work was thus concerned with internal evidences, drawn almost exclusively from the design argument. Paley's utilitarian ethics, his rational theology, and theologized natural history were required reading up until the 1840s and 1850s. At Oxford, too, the works of John Locke and Joseph Butler were used as texts throughout the eighteenth century, along with writings on natural theology such as Samuel Clarke's Boyle Lectures (1704–5), John

Wilkins's *Principles and Duties of Natural Religion* (1675), George Cheyne's *Philosophical Principles of Religion* (1715), and Robert Jenkin's *Reasonableness and Certainty of the Christian Religion* (1700).[14] The university curriculum in England thus reflected a fundamental conviction of the age—the unity of theological and physical truth.[15] Paley himself wrote of "the uniformity of plan observable in the universe," maintaining that the whole universe "is a system; each part either depending on other parts, or being connected with other parts by some common law of motion, or by the presence of some common substance."[16]

While Paley's *Natural Theology* may have set the tone, for the mid-nineteenth century it was the eight Bridgewater Treatises (1833-36) that collectively displayed how the various sciences shared this common integrating principle. On his death in 1829, the eighth Earl of Bridgewater made provision in his will for a series of works on "the Power, Wisdom, and Goodness of God, as manifested in the Creation." The president of the Royal Society, charged with the selection of appropriate authors, commissioned some of the most eminent scientists and philosophers of the period, including William Whewell, Charles Bell, William Buckland, and Peter Mark Roget. In the Preface to his contribution, Roget, a professor of physiology best known today as the author of *Roget's Thesaurus,* wrote that with the single principle of design, he had been able to marshal the facts of natural history into a "methodized order," to unite them into "comprehensive generalizations," and to establish their "mutual connections." He was thus able to bring "a unity of design, and that scientific form, which are generally wanting in books professedly treating of Natural Theology" while, at the same time, cherishing the hope that his compendium "might prove a useful introduction to the study of Natural History."[17]

Natural theology represented not only a substantive unifying principle for those investigating nature, but also provided a pedagogical orientation to those seeking to communicate truths about the natural world. The educational reformer Henry Brougham insisted that without "the sublime truths of natural theology . . . science cannot be well, any more than fairly taught."[18] A further advantage of the prominence of moral and religious themes was that the teaching of natural topics could be accommodated to the traditional moral goals of education. As one writer expressed it:

"God has arranged all things for the purpose of teaching us these [moral] lessons, and he has created our intellectual and moral natures expressly for the purpose of learning them."[19] At Cambridge, one of the founders of modern geology, Adam Sedgwick, regarded the discipline as an element of the liberal arts education, and in his examination questions asked students to reflect upon the philosophical implications of geology.[20] John Stevens Henslow, friend of Sedgwick and mentor of Charles Darwin, expressed similar hopes in his advocacy of university examinations in botany: "I rejoice to find the Natural Sciences at length taking firmer root in the seat of sound learning and religious education than they have hitherto obtained."[21] Indeed, when moral philosophy finally disappeared from the core university curriculum in the twentieth century, many university administrators cherished the hope that the inculcation of values would be taken over by the natural sciences.[22] Programs for the popularization of science were motivated by similar assumptions. John Brooke and Geoffrey Cantor note that many in the nineteenth century cherished the belief that "if the working classes were given a scientific education they would be drawn out of shallowness and sensuality into placid and upright citizens."[23] This sentiment received tangible and enduring architectural expression in the grand scientific edifices of the Victorian era. The scientific museum was intended to serve "as an institution in which the working classes—provided that they dressed nicely and curbed any tendency towards unseemly conduct—might be exposed to the improving influence of the middle classes."[24]

Allied to the idea that natural theology was integral to scientific endeavor was the persistent belief that scientific practice was morally ennobling for those involved in it. The assumption that the study of nature would promote moral qualities in its practitioners was, as we have seen, part of the classical legacy. In spite of the demise of the specific intellectual virtues in the seventeenth century, the idea that the study of nature contributed to a broad moral formation endured. Joseph Glanvill had declared in 1676 that the experimental approach would promote virtue and happiness in those who practiced it. In extolling the virtues of the nascent Royal Society, Thomas Sprat also insisted that the experimental philosophy would conduce to the composing and purifying of men's thoughts.

The natural philosopher and Minim friar Marin Mersenne wrote that the mathematical sciences "elevate the mind above itself" leading us to acknowledge the Deity and to imitate his works.[25] Sentiments such as these persisted until well into the nineteenth century. In his influential *Preliminary Discourse to the Study of Natural Philosophy* (1830) the astronomer and mathematician John Herschel wrote that the natural philosopher "is led to the conception of a Power and Intelligence superior to his own, and adequate to the production and maintenance of all that he sees in nature." Not only did nature provide evidence of the existence of the Deity, but it also had an inevitable effect on the character of the natural philosopher. Echoing the distant sentiments of Plato and Ptolemy, Herschel declared that "the observation of the calm, energetic regularity of nature, the immense scale of her operations, and the certainly with which her ends are attained, tends, irresistibly, to tranquilize and re-assure the mind, and render it less accessible to repining, selfish, and turbulent emotions."[26] In a similar vein, the scriptural geologist George Fairholme wrote in 1833 that the "great end of the study of Geology ought to be, a *moral* rather than a *scientific* one."[27]

From a twenty-first-century perspective, all of this looks like a rather odd amalgam of the physical, ethical, and theological—natural science mixed up with moral edification and pious sentiments about the Creator. But for the historical actors this apparent aggregation was the natural outcome of the presumed unity of knowledge. As John Herschel put it, "Truth is single."[28] It need hardly be pointed out that while this principle brought a unity to the sciences, broadly conceived, it also meant that truths about the natural world could not, in principle, be in conflict with religious truths. William Kirby wrote in his Bridgewater Treatise that the truths of nature and of revelation "cannot, if rightly interpreted, contradict each other, but must mutually illustrate and confirm." Whewell, who authored the third Bridgewater Treatise, wrote that "all truths must be consistent with all other truths, and . . . therefore the results of the geology or astronomy cannot be irreconcilable with the statements of true theology."[29] Considerations such as these led historian Robert Young to observe that the first six decades of the nineteenth century were characterized by "a common intellectual context" or what we would now call "a rich interdisciplinary culture" that was held together by a general natural theology. While historians have now demonstrated that the situation was rather more complicated than that, it is nonetheless true that natural theology provided a key resource for the communication of science to the general public, offered assurances that science could deliver a range of social and moral goods, and acted as an indispensable unifying principle for natural knowledge.[30]

"Science" and the Beatification of Bacon

Up to the middle of the nineteenth century, while many elements of natural history and natural philosophy could be regarded as "science" in our sense, the framework in which they were presented and the motivations for pursuing them were clearly rather different. As we have already noted, natural theology was itself numbered among the inductive sciences.[31] One writer went so far as to claim that natural theology was a certain science — "open to no objection," "in strict conformity with the rules of the inductive philosophy," and "consistently denied by those only who reject the 'Principia' of Newton."[32] If in the late seventeenth century it had been claimed that natural philosophy was a religious activity, in the nineteenth century it was suggested that natural theology was a scientific activity:

> Natural theology is an inductive science. Our knowledge of the existence of God, as far as that knowledge is traceable by the light of nature, is acquired by an intellectual process strictly analogous, and exactly similar, to the intellectual process by which we acquire our knowledge of the laws of the physical world. . . . Newton discovered the true system of the heavens; and it is only by this reasoning that the theist can ascertain, from the light of Nature, the existence and attributes of Him who made the heavens. The proof of a divine intelligence ruling over the universe, is as full and as perfect as the proof that gravitation extends throughout the planetary system.

It is significant that for the author of this 1843 article in the *Quarterly Review* not only was natural theology a science, it was an *inductive* science (which is to say that it was based on generalizations drawn from a large number of specific experiences). Natural theology was nothing other than "inductive philosophy . . . applied to theology." The proper inductive procedure thus ruled out "metaphysical arguments from first causes" and admitted only "facts ascertained by experience." Paley's only shortcoming, according to this author, was that "he was not a perfect master of the inductive logic."[33]

The method of induction, then, came to be thought of as a divinely sanctioned mode of reasoning that characterized both true religion and genuine science. The American Calvinist theologian Leonard Woods wrote in 1822 that theologians should regulate themselves "by the maxims of Bacon and Newton." The principles of observation and experience, he insisted, are "as applicable to the science of *Theology*, as to that of physics."[34] Princeton theologian Charles Hodge (1797–1878) agreed that Protestant theology conformed to the principles of Baconian science, insofar as it consisted in the drawing of generalizations from the assortment of facts contained in the Bible. According to Hodge: "The true method of theology is . . . the inductive, which assumes that the Bible contains all the facts or truths which form the contents of theology, just as the facts of nature are the contents of the natural sciences." Hodge's contention that the inductive method should be applied to the content of scripture was surprisingly common among evangelical writers. In his *Organon of Scripture* (1860), J. S. Lamar urged the application of Baconian methods to the text of the Bible. What Protestant exegetes desperately needed, in Lamar's view, was a method that gave a "scientific character" to biblical interpretation. Lamar insisted that there was nothing particularly novel about his approach: it "merely adopts, and applies to the Scriptures, a method which has been satisfactorily tried in other departments of study."[35] That method he associated primarily with Bacon, but also with John Herschel and J. S. Mill. This enthusiasm for Baconianism was not a solely American fad. Joseph Angus, principal of Regent's Park Baptist College during the 1860s, held that the whole of theology was an inductive science consisting of generalizations drawn from the facts of scripture. The methods of theology, he suggested, were directly analogous to those set out by Bacon.[36]

Many Protestants closely associated Baconian methods with their own theological allegiances, so much so that one commentator has spoken of the "beatification of Bacon" by theological conservatives during this period.[37] Presbyterian philosopher Samuel Tyler contended that "Protestant Christianity and Baconian philosophy originate in the same foundation" (something that Bacon himself had suggested some two hundred years earlier). James Marsh agreed that "by most persons they are considered as necessary parts of the same system."[38] That system was understood as necessarily opposed to more "speculative" approaches that characterized bad science and defective theology. Charles Hodge wrote that "as natural science was a chaos until the principle of induction was admitted and faithfully carried out, so theology is a jumble of human speculations, not worth a straw, when men refuse to apply the same principle to the study of the Word of God."[39]

The idea of natural theology as an inductive science served to reinforce the modern propositionally driven understanding of religion. On this understanding, "religion" itself was argued to be the product of induction—that is to say, a system of knowledge supported by a particular form of reasoning. William Paley had regarded religion as simply another "subject of human reasoning," a "system of divinity . . . founded on moral evidence."[40] The business of religion thus called for an ongoing investigation of the natural order. As Paley explained, the argument was cumulative. While each instance of contrivance was in itself a complete argument for divine agency, taken together, all such arguments had a more powerful effect than when considered singly.[41] Just as the cumulative weight of observation in the empirical realm was thought to lead to the formulation of ever more certain laws of nature, so too, as an anonymous reviewer of yet another nineteenth-century work on evidences expressed it, "the evidence of religion is of this cumulative kind."[42]

The idea of mental habits was not entirely lost in these discussions. William Whewell, hinting at the atheistic tendencies of certain of the French natural philosophers, insisted that the inductive mindset, in contrast to deductive habits of mind, was predisposed to arrive a theological conclusion: "I have already ventured to express an opinion that Inductive Minds, those which have been able to discover Laws of Nature, have also commonly been ready to believe in an Intelligent Author of Nature;

while Deductive Minds, those which have employed themselves in tracing the consequences of Laws discovered by others, have been willing to rest in Laws, without looking beyond to an Author of Laws."[43] Whewell went so far as to suggest that the success of the inductive method, along with the progress of natural knowledge that resulted from it, was in itself evidence of the providential designs of the Deity.[44] Whewell even dreamed of founding an inductive school that would extend its reach to moral philosophy, political economy, and science.[45]

Other authors of the Bridgewater Treatises expressed similar views. Peter Roget observed: "It is by pursuing the method of philosophical induction, so well explained by Bacon, that the physical sciences, which the misdirected efforts of former ages had failed to advance, have, within the last two centuries, been carried to a height of perfection which affords just grounds for exultation in the achievements of the human intellect."[46] William Kirby, the "father of entomology," also paid homage to "the great and wonderful genius . . . Lord Veralum [i.e. Francis Bacon], who laid the foundation upon which the proud structure of modern philosophy is erected." "For the discovery of physical truth," Kirby insisted, "we must have recourse to induction from experiment and soberly conducted investigation of physical phenomena." This sober method of proceeding, which in his view would ineluctably lead to a theological conclusion, was to be contrasted with the rash speculations of such naturalists as Laplace and Lamarck, whose hypothesizing amounted to nothing other than "self worship . . . self delusion, and the love of hypothesis."[47] Whewell contended that one of the lessons to be drawn from the history of science was that there was a strong correlation between method, moral character, and outcomes. Thus Descartes was "rash and cowardly," his hypothetical method presumptuous, and his theories fallacious.[48]

This identification of two distinct modes of scientific reasoning—the inductive (good) and the deductive (not so good)—enabled Whewell to account for apparent instances of conflict between religion and knowledge of nature. Such conflict arose when the relevant "science" arose out of deductive and speculative methods, and was the product of minds insufficiently shaped by the sober methods of induction. Related to this was the fact that speculation divorced from familiarity with the facts was conducive to atheism, which in turn led to social instability. According to Edmund Burke's (1729–97) influential analysis, an unholy alliance of atheism and materialism had provided ideological succor for the forces of revolution in France.[49] A motive for maintaining the synthesis between empirical science and natural theology, and for the preaching of properly inductive sciences to the working classes, was the hope that they would be thereby rendered pious and upright citizens, disinclined to foment rebellion.[50] Whewell thus attributed England's political stability to the powerful physico-theological synthesis preached there.[51]

Later in the century these moral and methodological considerations were to play an important role in the controversies generated by the theories of evolution. Sedgwick wrote in his attack on Robert Chambers's *Vestiges of the Natural History of Creation* (1844)—a best-selling work that argued for the progressive transmutation of species—of his cherished hope "that human knowledge will go on in the right road of sober Inductive Truth" and that by beginning with the humble contemplation of "the order of nature . . . may rise, step by step, to a more lofty knowledge."[52] Contemplation of nature was thus a prelude to the consideration of higher moral and religious truths. But such contemplation was to be guided by inductive logic and not wild speculation. Subsequently a number of Darwin's critics would chide him for having departed from the path of sober induction.[53]

Praise of the Baconian method could extend to the view that inductive habits of mind not only promoted the cause of true science, but also provided the basis of proper moral and religious sentiments. As we have seen, such views were not uncommon in the seventeenth century, when proponents of experimental natural philosophy insisted that its practice would induce sound habits of mind.[54] Indeed, it was argued that there was a strong affinity between experimental science and what was then commonly called "experimental religion"—a religion based on experience and trials. Philosophical theologian Jonathan Edwards (1703–58) explained: "As that is called experimental philosophy, which brings opinions and notions to the test of fact; so is that properly called experimental religion, which brings religious affections and intentions to the like test."[55] Likewise, while Paley's *Natural Theology* is typically read as an extended

inductive *argument,* he, too, reflected on the way in which the attempt to
draw together a common pattern from particulars was virtually a form of
religious *meditation:*

> Now it is by frequent or continued meditation upon a subject, by placing
> a subject in different points of view, by induction of particulars, by variety
> of examples, by applying principles to the solution of phenomena, by
> dwelling upon proofs and consequences, that mental exercise is drawn
> into any particular channel. . . . In a *moral view,* I shall not, I believe, be
> contradicted when I say, that, if one train of thinking be more desirable
> than another, it is that which regards the phenomena of nature with a
> constant reference to a supreme intelligent Author. To have made this
> ruling, the habitual sentiment of our minds, is to have laid the foundation
> of everything which is religious. The world thenceforth becomes a temple,
> and life itself is one continued act of adoration.[56]

Other eighteenth-century writers would endorse this understanding.
Thomas Chalmers, one of the Bridgewater authors, contended that there
was a "close affinity between the taste for science, and a taste for sacred-
ness." The systematic study of nature could "make man a more reflective
and less sensual being."[57] The edifying properties of inductive methods
thus bring us again to the idea of the natural philosopher as being possessed of special virtues.

In sum, although the intellectual virtue *scientia* may have been dispensed with in the seventeenth century, it was still assumed that certain
habits of mind—closely allied with moral sensibilities—were necessary
for the correct prosecution of science. The difference now was that the
speculative and deductive habits of mind associated with *scientia* were displaced by inductive and experimental dispositions. This meant that some
systems of natural knowledge—those premised on speculation alone—
had the capacity to stand in opposition to religion. But true scientific reasoning, based on induction, could never yield such conflict. In the English
context, the scientific mindset, almost by definition, was understood to be
necessarily favorable to religion. More than this, inductive habits of mind
were thought to shape the character of men of science, equipping them

morally for the proper investigation of nature. By the turn of the twentieth century, however, searching questions were being asked about the moral qualifications of the scientist and whether a scientific mentality was necessarily compatible with religious sensibilities.

The Scientist

In a famous essay, "The Will to Believe," first delivered in 1896 as a lecture to the philosophy clubs of Yale and Brown universities, the American philosopher and psychologist William James (1842–1910) broached the issue of the moral basis of science. He observed that science was "absolutely impersonal" and indifferent to the moral lives of its exponents. James would not be alone in this view. In an equally celebrated 1918 lecture, "Science as a Vocation," sociologist Max Weber followed suit, dismissing the claim that science was morally edifying. According the Weber, this lofty ideal was a fiction perpetuated by "a few big children in university chairs." For good measure, Weber added that the sciences were also bereft of religious content: "that science [*Wissenschaft*] today is irreligious no one will doubt even in his innermost being."[58] Science, for James and Weber, was a disinterested and agnostic activity that demanded neither special moral qualities nor religious commitments of its practitioners.

Assuming that these characterizations are correct—and they are largely consistent with the present view of the natural sciences—the obvious question to ask is this: what happened to the moral and theological elements that had once been integral to the study of nature and the person of the naturalist? The short answer is that the second half of the nineteenth century witnesses the disintegration of the common religious and moral context of scientific endeavors, and sees the reconstruction of "science" around the principle of a common method and a common identity for its practitioners. At the same time, the older moral and theological unifying principles are systematically rejected as this new aggregation seeks actively to demarcate its new territory and to distinguish itself from a range of newly "nonscientific" activities. Modern science, then, emerges from a threefold process: first, a new identity—the scientist—is forged for its practitioners; second, it is claimed that the sciences share a distinc-

tive method, one that excludes reference to religious and moral considerations; and, third, following on from this, the character of this new science
is consolidated by drawing sharp boundaries and positing the existence of
contrast cases—science and pseudo-science, science and technology, science and the humanities and, most important for our purposes, science
and religion. This last development was accompanied by the construction
of a mythical past in which the newly crafted boundaries of the disciplines
assume an ahistorical timelessness, and tales of a perennial conflict between science and religion are fashioned to reinforce the contemporary
lines of demarcation.

The emergence of the idea of "the scientist" is familiar territory to historians of science, but it is worth repeating key elements of it here. Writing in an 1834 review of Mary Sommerville's *On the Connexion of the Physical Sciences*, William Whewell took the opportunity to rue the centrifugal
tendencies of the science of nature, complaining that the soil of the natural sciences was being divided into "infinitely small allotments." Natural
science, he declared, was like "a great empire falling to pieces."[59] Whewell
was by no means alone in his assessment. In 1818, philosopher and Roman-
tic poet Samuel Taylor Coleridge had already complained that "the terms
system, method, science are mere improprieties of courtesy when applied
to a mass enlarging by endless oppositions, but without a nerve that oscil-
lates, or a pulse that throbs."[60] "The natural sciences have remained devoid of mutual connexion," agreed one reviewer in 1837, before going on
to suggest that the declining role of philosophy as a unifying principle was
at least partly to blame.[61] These judgments suggest that as the nineteenth
century progressed, the "common intellectual context" provided by natural theology was being threatened by an increasing specialization of the
natural sciences and the narrow focus of many of their practitioners.

Part of the problem, Whewell believed, was "the want of any name by
which we can designate the students of the knowledge of the material world
collectively."[62] In an attempt to remedy this deficiency, he coined the term
"scientist," which, of a range of suggested alternatives—"philosophers,"
"savans," "nature poker," "nature peeper"—he thought the most dignified.
Not all agreed with Whewell's choice, it must be said, and while he had
sought some lofty analogy with the term "artist," for others the term was

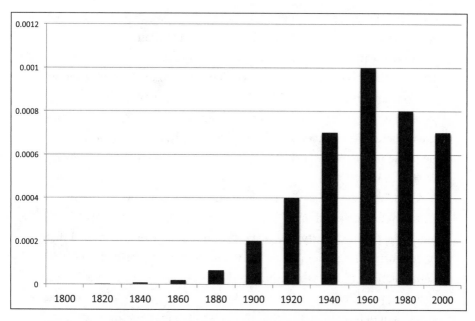

Figure 9. Percentage of English books containing the term "scientist," 1800–2000.

redolent of "dentist," one engaged in the paid performance of an unpleasant profession. The suspicion that the word had originated in America, although untrue, also counted against its adoption. Alexander Ellis, fellow of the Royal Society and president of the philological society, pompously declared "scientist" to be "an American barbarous trisyllable."[63] Indeed, it was true that across the Atlantic the term had received a warmer welcome. Perhaps what ensured its eventual triumph in England was the difficulty in pronouncing the term that was its chief rival—"physicist."

The adoption of this label was facilitated by the appearance of new bodies for scientists, and by the increasing professionalization of existing ones. The British Association for the Advancement of Science was established in 1831 and sought "to give a stronger impulse and a more systematic direction to scientific inquiry." One of its chief aims was to counter the effects of increasing specialization by seeing what light the various sciences were "capable of mutually reflecting upon each other."[64] The Royal Society, whose fellowship had been dominated by gentleman amateurs, tightened its membership rules in 1847, seeking to ensure that those elected had genuine scientific interests and accomplishments. One prominent feature of the changing membership profile of these organizations was the decreasing proportion of Anglican divines in both executive positions and in the rank and file. In the wake of the reform of the Royal Society's membership rules, the fifty years from 1849 to 1899 saw an absolute reduction in members, but more significantly, an overall decrease in the participation of Anglican clergymen from 9.7 percent to 3.1 percent (see figure 10).

A similar pattern of clergy involvement may be seen in the British Association. In the period 1831–65, nine clergymen acted in the role of president; over the next thirty-five years, 1866–1900, no clergyman held that office. During the earlier period, forty-one clergymen presided over individual sections of the association; in the latter period, only three did so. Declining participation of the clergy is also evident in the informal local societies of natural history, although the trend tends to occur much later. In 1909, the Lincolnshire Naturalists had ninety-seven members, of whom 20 percent were clergymen. By 1940 this representation had been reduced to 3.3 percent.[65]

Diminishing clerical membership in scientific societies was one of the consequences of a growing specialization and professionalization of the sciences. But it was also partly the outcome of a deliberate attempt to reduce clerical influence on the political machinery of scientific organizations. This was the explicit mission of Thomas Henry Huxley and his colleagues in the "X-club"—a group not inaccurately described as a kind of "masonic Darwinian Lodge."[66] Formed in 1864, the club's membership was united by a "devotion to science, pure and free untrammelled by religious dogmas."[67] The group met almost monthly for a thirty-year period, and one of their goals was to establish a secular network that would serve as an alternative to the clerically dominated scientific hierarchy. Members were politically active, successfully contriving to alter the election rules of the Royal Society to have their allies elected and extending their influence even to the position of the presidency. In a major coup, they managed to have Darwin awarded the society's most prestigious prize, the Copley Medal, despite determined opposition and an extremely close vote.[68]

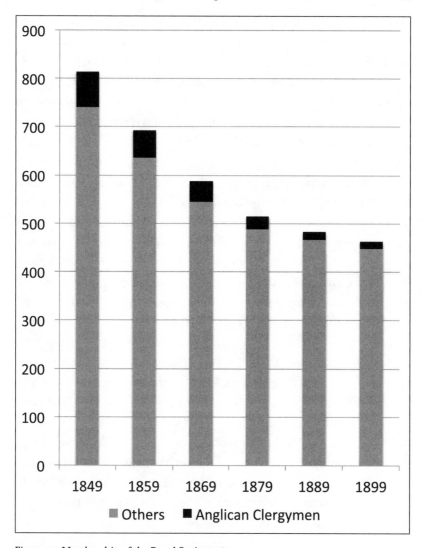

Figure 10. Membership of the Royal Society, 1849–99.

These individuals also sought with an evangelical fervor to establish a scientific status for natural history, to rid the discipline of women, amateurs, and parsons, and to place a secular science into the center of cultural life in Victorian England.[69]

The professionalization of science necessarily brought with it a reappraisal of the moral and religious qualifications thought to be essential to the investigator of nature. Francis Galton, whose *English Men of Science* (1874) sought to investigate the personality of scientists and determine the roles played by heredity and environment in their development, concluded that the offices of priest and scientist were incompatible. Relying upon his analysis to add to the polemic of the scientific professionalizers, he observed that "the sons of clergymen rarely take a lead in science" and that "the pursuit of science is uncongenial to the priestly character." The justification for this claim lay in Galton's observations of the declining participation of the clergy in scientific activities. He contended that of the 660 council posts in scientific societies from 1850 to 1870, only sixteen (2.4 percent) had been occupied by clergyman, mostly astronomers, mathematicians, and physicists with "not a single biologist among them."[70] As for his assertions about the incompatibility of "the priestly character" and scientific endeavor, these represent a remarkable reversal of the standard assumptions of the previous generation of men of science, many of whom regarded piety as an almost essential prerequisite for facility in the study of nature. This view of the scientific vocation is as remote as it could be from the idea of "priests of nature," advocated by Robert Boyle and a number of his contemporaries (although Galton and Huxley continued to use such terms as "scientific priesthood," "professorial episcopate," and "church scientific" for their scientific fraternities). More significantly, it represents a clear breach with the cherished assumptions of influential commentators on science, such as William Whewell, and a decisive step in the direction of the amoral scientific persona described by William James and Max Weber.

Science and Its Methods

The changing character of those engaged in scientific endeavors was matched by a new nomenclature for their endeavors. The most conspicuous marker of this change was the replacement of "natural philosophy" by "natural science." In 1800 few had spoken of the "natural sciences" but by 1880, this expression had overtaken the traditional label "natural philosophy." The persistence of "natural philosophy" in the twentieth century is

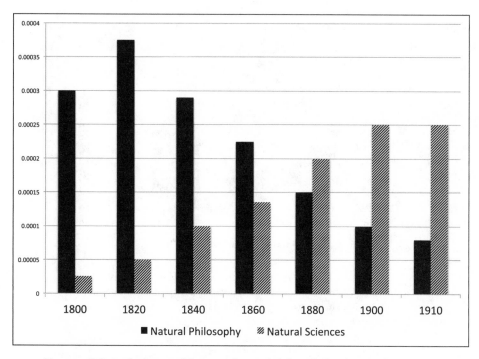

Figure 11. Relative frequency of the terms "natural philosophy" and "natural sciences" in English books, 1800–1910.

owing largely to historical references to a past practice (see figure 11). As should now be apparent, this was not simply the substitution of one term by another, but involved the jettisoning of a range of personal qualities relating to the conduct of philosophy and the living of the philosophical life. It is noteworthy that Whewell had actually considered the generic term "philosopher" as an alternative to "scientist," but had dismissed it as "too lofty." In any case, as can be seen from William Ward's subsequent and more restrictive definition of science, "philosopher" was no longer an appropriate label, given that theological and metaphysical concerns now lay beyond its disciplinary boundaries.

Natural history suffered a similar fate to natural philosophy, with the relatively novel "biology" coming to stand in for a more "scientific" study of life. The term "biology" appears at around the turn of the eighteenth

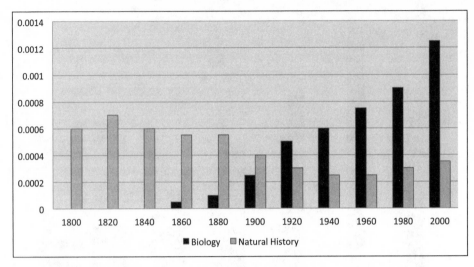

Figure 12. Relative frequency of "natural history" and "biology" in English books, 1800–2000. "Biology" was taken to be the "scientific" discipline, while natural history was thought to be the preserve of amateurs, parsons, and women.

century. In the very first English occurrence of the term in 1799 we encounter the physician Thomas Beddoes declaring that "biology" — the doctrine of living systems — is "the foundation of ethics and pneumatology." (By "pneumatology" he probably means the science of the human mind, although more generally this referred to the study of spirits and spiritual beings.) Beddoes maintained that such knowledge was a necessary prerequisite to "progress in genuine morality."[71] Initially, biology retained some of the moral connotations of the natural history it was destined to replace, but these gradually ebbed away as biology was incorporated into the new rubric of science. Other early references to "biology" are translations of "biologie," a term used by both Jean-Baptiste Lamarck in Hydrogéologie (1802) and Gottfried Reinhold Treviranus in Biologie oder Philosophie der lebenden Natur (1802) (see figure 12).[72] The association of the term with Lamarck (who was widely regarded as exemplifying an impious and overly speculative approach to nature) was something of an obstacle to its early adoption, but by mid-century Whewell could observe that the word "has of late become not uncommon, among good writers."[73]

Biology came to be thought of as a more scientific approach to the natural world for a number of reasons. First, as the name suggests, natural history entailed a "historical" approach to the study of living things, with descriptions of organisms bearing some comparison to descriptions of human deeds in the discipline of civil history. In the early modern period, there were understood to be two branches of history—natural and civil—and both had been given a didactic, moral function. In other words, the record of human deeds *and* the descriptions of living things had an exemplary function, and this was carried over into the physico-theology of the eighteenth century. The adoption of the new term "biology" brought with it the opportunity to divest the study of nature of these lingering, but nonetheless persistent, exemplary elements.

Two further reasons for the scientific status of biology were to do with the identity of its practitioners and the site of their activities. Those strongly committed to the scientific study of nature could now identify themselves as "biologists." This came to be regarded as a much more serious occupation than that of "naturalist." T. H. Huxley explained that "the word 'naturalist' unfortunately includes a far lower order of men than chemist, physicist, or mathematician . . . every fool who can make bad species and worse genera is a 'Naturalist.'" For Huxley, the problem with naturalists was that they were insufficiently scientific and that anyone could call himself or herself a naturalist.[74] Furthermore, the growth of experimental biology brought with it a new venue for the biological research—the laboratory. And whereas the natural world had been available to all, the restricted nature of laboratory access necessarily meant that amateurs were excluded from making contributions to the new and increasingly specialized experimental biology.[75]

When we consider the fate of these two terms, "natural history" and its twin "natural philosophy," it is worth recalling that they bear, in their very nomenclature, genealogical connections to the humanistic disciplines of philosophy and history. The gradual replacement of these two terms by a generic "science" and the host of subdisciplinary specializations that included "biology" represents a significant severing of ties with the humanities and the ominous bifurcation of Western intellectual life into what C. P. Snow, in the 1950s, would famously designate "the two cultures."

If the invention of the word "scientist" helped mark out science as a distinctive and unified activity, an equally important mechanism for lending coherence to "science" was the idea of *the* scientific method. At first sight, it may seem that "the inductive method" would serve the purpose of giving a methodological unity to the new notion of the natural sciences. But one problem with invoking induction as a unifying method was that it was insufficiently exclusionary. The inductive method, as we have seen, was widely claimed to characterize the methods of theology, both natural and revealed. Some argued that ethics, too, stood on an inductive foundation. The idea of a science that excluded the philosophical and theological required recourse to a more narrow understanding of "the scientific method" that could unite the increasingly specialized natural or physical sciences, while at the same time excluding the theological and the metaphysical.

In keeping with the indiscriminate uses of the term "science" in the first half of the nineteenth century, talk of *a* scientific method had initially meant simply a systematic plan of attack that could be applied to any number of activities, from physiology to fishing. But from the 1870s onward we see a growing number of references to *the* scientific method that is far more restricted in application, and is associated with the practices of the new profession of the scientist (see figure 13). These more technical usages replace not only generic understandings of *a* scientific method, that were applicable to virtually any activity, but also stand in for once common references to the Baconian method, or Baconian induction. In his 1874 *Principles of Science: A Treatise on Logic and the Scientific Method*, the economist and philosopher William Stanley Jevons thus took against the purely empirical methods of Bacon, offering an account of the scientific method that sought to capture the methods actually employed by Newton, Huygens, and Faraday.[76] In a further indication that the tide was turning against Baconian inductivism, the 1875 edition of the *Encyclopaedia Britannica* came out against a purely inductive approach to science:

> The inductive formation of axioms by a gradually ascending scale is a
> route which no science ever followed, and by which no science could ever
> make progress. The true scientific procedure is by hypothesis followed up

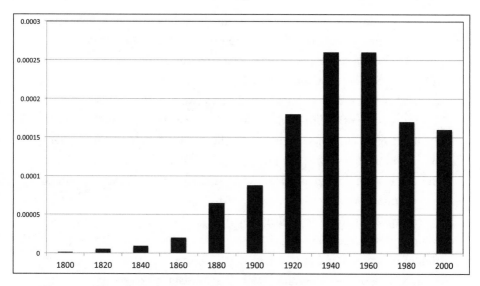

Figure 13. "Scientific method" in English books, 1800–2000.

and tested by verification; the most powerful instrument is the deductive
method, which Bacon can hardly be said to have recognised.[77]

No longer did nineteenth-century experimentalists catalog observations
in the hope that a natural pattern would emerge unbidden; rather, they
specifically sought phenomena relevant to a hypothesis and designed experiments with a view to deciding between competing theories.[78] These
changes in turn reflect a new conception of the sciences. The sciences of
nature did not entail the humble accumulation of data in the hope that
nature would reveal its secrets. Rather, order is imposed by the human observer. Science becomes fundamentally a human activity, and the major
player is the scientist. Prior to this, nature itself was the focus, and the
chief player, albeit one who transcended physical phenomena, was God.
Now the wonders of nature became the wonders of science, understood
as the product of scientists' rigorous application of the scientific method.

While there was an explicit attempt to establish the exclusivity of science, at the same time claims were made for the wide applicability of its
methods. Thomas Henry Huxley thus reported in 1871 that "all men of

science would more or less pin their faith" on the proposition that "the scientific method was the only method by which truth could be ascertained." This, he suggested, was a necessary implication of the idea of the ultimate unity of truth, for "unless anyone was prepared to maintain that there were two different sorts of truth and two different kinds of Logic, then it was inevitable that what they called the scientific method must extend itself to all forms of enquiry."[79] The scientific method, thus understood, entails the view that underlying the various scientific disciplines there is a single unified and generic "science," and that this science offers us a unique and privileged access to truth.

The appearance of the scientific method, along with the loss of the terminology of "natural philosophy," also marks the final stage of the evacuation of personal moral qualities from the persona of the man of science. Taking stock of the common view "that scientists are recruited from the ranks of those who exhibit an unusual degree of moral integrity," sociologist Robert K. Merton could find no satisfactory evidence to support it. Rather, he concluded, any putative personal virtues are a function of a "distinctive pattern of institutional control." It is not the moral qualities of its practitioners that imbue science with its particular ethos, but rather the institutional conventions of science, embodied in what are thought to be its distinctive methods.[80]

Together, then, the creation of a special professional identity (the scientist), the specification of a distinguishing set of methods (the scientific method), and the replacement of a traditional nomenclature (natural philosophy and natural history), signal the distillation of modern science out of a range of social practices in which it had been embedded. This coalescence of science in the late nineteenth century makes possible for the first time a relationship of science with religion, which had itself undergone a similar process of disembedding some centuries before. These developments make it possible to speak about science and religion as if they were two independent things that had a real existence in the world.

Boundary Work and Myth Making

When did people first begin to speak about science and religion, using that precise terminology? As should now be apparent, this could not have been before the nineteenth century. When we consult written works for actual occurrences of the conjunction "science and religion" or "religion and science" in English publications, that is exactly what we discover (see figure 14).[81] Admittedly, some of these usages—and particularly those in the first half of the nineteenth century—reflect the much broader understandings of "science" that we have already considered. In Nicholas Wiseman's *Lectures on the Connexion between Science and Revealed Religion* (1836) the "sciences" in question are philology, physiology, geology, chronology, archaeology, and biblical criticism.[82] Wiseman's volume is really not about "science and religion" in any sense that would be recognizable now. From the middle decades of the century, however, discussions of science and religion begin to take on a familiar complexion. With increas-

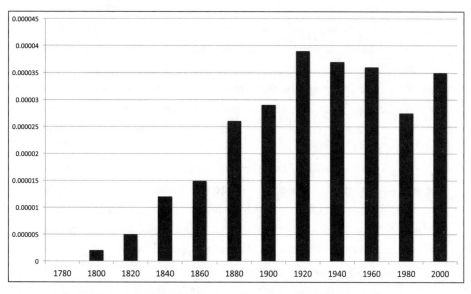

Figure 14. Percentage of English books containing the expression "science and religion," 1780–2000. The pattern for "religion and science" is virtually identical, although overall the former expression is significantly more common.

ing frequency, moreover, the relationship is described as one of conflict. Furthermore, when historical perspectives are introduced, the assumption is that religion and science are long-standing features of Western society and that the progressive course of history can be understood in terms of their perennial and conflicted relationship. The emergence of the modern notion of science was thus accompanied by a distorting projection back in time of this recently emerged conception. This amounted to the construction of a history that served the purposes of reinforcing recently emerged boundaries.

This story of the invention of the "conflict myth" is now a commonplace among historians of science, but it is a story that bears retelling. Two of the chief architects of the myth were John William Draper (1811–82) and Andrew Dickson White (1832–1918). Draper was an American chemist and amateur historian who in 1875 produced the seminal *History of the Conflict between Religion and Science*. White, the inaugural president of Cornell University, followed up with a two-volume *History of the Warfare of Science with Theology in Christendom* (1897). Between them, Draper and White cataloged or constructed most of the now prevalent myths about the putatively pernicious influence of religion on science—medieval Christianity was inhospitable to science, the church banned human dissection, Pope Callixtus III had excommunicated Halley's comet, ancient and medieval thinkers believed in a flat earth, heliocentrism was resisted because it demoted humans from the center of the cosmos, Galileo was tortured and imprisoned by the Inquisition, the church opposed vaccination and anesthesia, Darwin's *Origin* was met with universal resistance from religious quarters, and so on.[83]

A significant number of these episodes are sheer fabrications—the flat earth myth being perhaps the most egregious example. Others have an undeniable historical foundation, but are constructed in ways that would have utterly perplexed the original actors. There is no doubt, for instance, that Galileo was tried by the Inquisition and forced to recant the Copernican hypothesis. But to cite this as an instance of science-religion conflict is to misconstrue the context. For a start, the Catholic Church endorsed the scientific consensus of the period, which, on the basis of the available evidence, held that the earth was stationary in the middle of the cosmos. To this extent it might be better to characterize the episode as a conflict *within* science (or, more strictly, within astronomy and natural philosophy) rather than between science and religion. Second, the first use of the Galileo affair for propaganda purposes was by Protestants seeking to discredit Catholics, so that it was initially given a role in conflicts *within* religion. Related to this is the fact that the Copernican hypothesis had first been postulated some eighty years before the trial of Galileo, and hence the context of the Protestant Reformation is a key to understanding why the papacy took steps at this particular time. Finally, even if it could be constructed as a science-religion conflict, the condemnation of Galileo was not typical of the Catholic Church's attitude toward the study of nature, since at the time the church was the single most prominent supporter of astronomical research.

But these are myths not only because they are historically dubious, but also because they fulfill a traditional function of myth—that of validating a particular view of reality and a set of social practices. This accounts for the persistence of these myths in spite of the best efforts of historians of science. At the same time, the myths also serve to reinforce the boundaries of religion, which has a negative role to play in shoring up the identity of modern science. It is significant, then, that both Draper and White insisted that science and religion were enduring and opposed features of Western history and of styles of thinking. "The history of science," Draper tells us, "is a narrative of the conflict of two contending powers." White spoke in similar terms of "a conflict between two epochs in the evolution of human thought—the theological and the scientific."[84] For Draper and White, the lack of an inherent unifying principle among the sciences is compensated for by the promotion of a kind of negative definition, in which science is understood by what it is not, or by what it is in opposition to—in this case, religion. This is akin to the old ploy of uniting a fractious nation by focusing attention on an external enemy. Just as the perpetual warfare waged by the fictional state Oceania in George Orwell's 1984 was crucial to that state's integrity, so the fictional historical warfare waged between science and religion served the contemporary need of consolidating the new boundaries of science and promoting internal cohesion.

The warfare metaphor was reinforced with another cherished

nineteenth-century notion—the idea of historical progress. On one common understanding of cultural evolution, religion represents a primitive phase of human development that is destined to be replaced by a more scientifically enlightened age. Draper, for example, contended that the state of Europe at the close of the Middle Ages offered an instructive example of the contributions of Catholic Christianity to the progress of human welfare; America, at the end of the nineteenth, exemplified the benefits of science.[85] The progress of human history toward the scientific age paralleled a similar story of progress in the organic realm: "From point to point in this vast progression there has been a gradual, a definite, a continuous unfolding, a resistless order of evolution. . . . By degrees, one species after another in succession more and more perfect arises, until, after many ages, a culmination is reached."[86] Again, then, "science" takes on a metaphysical significance far greater and more coherent than the disparate specialized practices from which it had been aggregated. It now becomes the *telos* or goal toward which society naturally progressed.

White's idea that science and theology represent two epochs of human thought also bears the unmistakable impress of the historicist vision propounded by the positivist philosopher Auguste Comte. As noted earlier, in his six-volume *Course on Positive Philosophy* (1830–42) Comte argued that the present age was witness to humanity's transition from the "metaphysical stage" to the higher "positive" or scientific level of development.[87] Theology, for its part, had preceded the metaphysical phase, and thus represented the most primitive form of cultural development. Such schemes of progress were to become commonplace in the late nineteenth century. In one of the seminal works of modern anthropology, *The Golden Bough* (1890), J. G. Frazer laid out a scheme of historical development in which the human race progresses from magic, through religion, to science.

While the conviction that a scientific age would necessarily follow a religious one might seem at odds with the image of an ongoing conflict, it was typically assumed that earlier episodes of conflict were triggered by exceptional individuals who had been born ahead of their time and who struggled against the forces of ignorance and superstition. Now their time had come. In the closing lines of his book Draper confronted his readers with a stark choice: "The time approaches when men must take their choice between quiescent, immobile faith and ever-advancing Science—faith, with its mediaeval consolations, Science, which is incessantly scattering its material blessings in the pathway of life, elevating the lot of man in the world, and unifying the human race. Its triumphs are solid and enduring." "Religion," he concludes, "must relinquish that imperious, that domineering position which she has so long maintained against Science."[88] And so it was that the nineteenth century oversaw a transfer of social prestige and power from religion to science. As historian (and member of the London Positivist Society) A. W. Benn observed of his own generation, "a great part of the reverence once given to priests and to their stories of an unseen universe has been transferred to the astronomer, the geologist, the physician, and the engineer."[89]

Diplomatic Relations?

The coalescing of the category "science" made possible the modern relationship between science and religion and the way in which the boundaries are presently understood delimits the range of their possible interactions. A common typology of science-religion relations sets out four such possibilities: conflict, independence, dialogue, and integration. This classification was first developed by Ian Barbour, a pioneer of the science-and-religion field, and while it is not uncontested, it remains highly influential.[90] As should by now be apparent, the "integration" category is one that applies almost exclusively to the pre-nineteenth-century period, when natural philosophy and natural history were closely integrated with natural theology. In a sense, then, this category does not refer to a mode of relations between science and religion at all, because to the extent that science has this intimate relationship with religion, it is not recognizably modern science. This relates to the general claim that I have been making throughout this book, that if we attempt a rigid application of our modern categories to the past, we are bound to arrive at a distorted picture. Turning to the present, what makes each of the remaining modes of relationship possible, along with the variety of standpoints within them, is the fact that neither category consistently picks out some discrete feature of the world. To put it another way, these labels do not identify anything resembling a natural kind.

When we consider what this means *within* the modern categories themselves, there is a telling difference between what follows for "religion" and what follows for "science." In the case of the former, the idea of a generic religion, of which the world religions are specific instances, gave rise to the problem of religious pluralism. Here, the idea is that because the world religions all make competing truth claims, at most only one can be true. But this conundrum is largely an artifact of the category: first, because of the historical development in which true religion comes to be understood in propositional terms; second, on account of the forced enlistment of the so-called world religions into this Western category. When we reflect on Augustine's rather different understanding of "true religion," or on Ficino's idea of "true Christian religion," we get a clear sense of what difference it makes to think in terms of the modern category.

As for science, while the process of reification is similar in some respects to that of religion, the differences are equally important. The modern construction of "science" was less an unintended consequence of other historical developments, and more an intentional project. As a consequence of this, the parties to the aggregation were typically willing and enthusiastic participants. While the category "religion" has a tendency to generate false tensions, the category "science" has the opposite effect, producing an artificial unity. The advantage of bringing together natural history and natural philosophy under the banner of "science" was that natural history (or biology, as it was to become) was able to lay claim to the best features of the practices of natural philosophy, and vice versa. This led to a consolidation of their respective strengths, resulting from a sharing of the more successful cognitive features of each of the constituent practices that went into the new "science." The mathematical rigor of the physical sciences thus combined with the matter-of-fact "historical" claims of the biological and geological sciences to give modern science an unassailable epistemic status. At the same time, the impressive technological achievements associated with the sciences (what Aristotle would have called the "productive" sciences) could be harnessed to lend authority to more ambitious theoretical claims about the world (what for Aristotle was the more elevated sphere of natural philosophy).

The present cognitive authority of science thus rests on a subtle conflation of the diverse goals of its once distinct constituents: the construction of empirically adequate models (mathematical astronomy), the production of useful technologies (arts or productive sciences), detailed descriptions of the natural world (natural history), and the provision of true causal accounts of the operations of nature (natural philosophy). The seventeenth century provided a necessary first step for the eventual consolidation of modern science, because it sought to unify mathematical astronomy (which offered empirically adequate models) and natural philosophy (which sought a single, true, causal account). The nineteenth-century construction of science completed the process by bringing about a unification of natural history and natural philosophy, while at the same time harnessing the practical benefits offered by technology to lend additional weight to its authority.

Turning to relations *between* science and religion, the various alternatives of conflict, independence, and dialogue arise out of the fact that both are considered to be predominantly concerned with knowledge. As a consequence, attention is focused either on the propositional content of the two enterprises or on the methods or modes of knowledge that generate those propositions. Whether the relations between science and religion, thus construed, are positive or negative depends to some degree on how the present boundaries of the categories are understood, and on the extent to which vestiges of their pre-nineteenth-century boundaries intrude into their present self-understandings.

A superficial analysis might seem to suggest that the most conspicuous contemporary conflict between science and religion—that between "creationists" and evolutionists—arises, on the one hand, out of an improper attempt on the part of religiously motivated antievolutionists to claim what is properly the territory of science and, on the other, out of an overreaction on the part of some spokesmen for science, who condemn all of religion on account of the infelicities of "scientific creationists." Here the role played by the categories themselves is clear, since the part (creationism/evolution theory) is taken for the whole (religion/science). But the evolution wars can conceal a more fundamental source of conflict, in which both science and religion compete for the more important territory that was once shared as part of the "common context"—when science is understood as having moral or religious implications and religion, conversely, as having scientific implications.

In some of its more popular guises, science is presented as possessing the philosophical or religious gravitas that had characterized natural philosophy and natural theology. Once, the cosmos had been a source of meaning and value. Now, following the consolidation of evolutionary theory in the first half of the twentieth century, it is nature viewed through an evolutionary lens that is most often conscripted for this role. Hints of these ambitions were already evident in the portentous closing lines of Darwin's *Origin*:

> There is grandeur in this view of life, with its several powers, having been originally breathed by the Creator into a few forms or into one; and that, whilst this planet has gone cycling on according to the fixed law of gravity, from so simple a beginning endless forms most beautiful and most wonderful have been, and are being, evolved.[91]

Acutely conscious of the intimate connection between nature and the moral order, Darwin had sought to replicate in his own narrative the kinds of moral and religious sentiments that were associated with natural history.

In the twenty-first century, attempts to imbue science with quasi-religious significance play little role in routine scientific activities, but are common in some influential popular presentations of science, particularly among those who seek to promote the image of an essential antagonism between science and religion. Harvard biologist E. O. Wilson maintains that scientific materialism "presents the human mind with an alternative mythology that until now has always point for point in zones of conflict, defeated traditional religion." Wilson goes on to say that the story of evolution—or, the "evolutionary epic" as he calls it—"is probably the best myth we will ever have."[92] Advocates of the new discipline of "big history," which seeks to unite the vast and disparate scales of cosmological, geological, and historical time, speak similarly of the epic of evolution and of a modern creation myth.[93] At times, these narratives are interwoven with the kind of moralizing homilies that had once characterized works of natural history. Wilson explains that science, after a long and arduous journey, was finally able to provide an alternative account of "man's place in the universe," relying upon "the scientific method." Of the sciences, it is biology that has become "foremost in relevance to the central questions of philosophy, aiming to explain the nature of mind and reality and the meaning of life." The moral imperative to which biology directs us is "stewardship of life."[94] This idea that science somehow directs us toward an environmental ethic has become commonplace. The American philosopher Loyal D. Rue thus agrees with Wilson that the epic of evolution produces "a new wisdom tradition that couples an evolutionary cosmology to an ecocentric morality."[95] Most recently, the "new atheists" have further advanced these general claims, arguing that the biological sciences provide the ultimate guide to life's most profound questions. Sam Harris has boldly insisted that questions "about meaning, morality, and life's larger purpose" are ultimately questions that science, and not religion, can answer. Science, he informs us, "will gradually encompass life's deepest questions." Richard Dawkins similarly claims that whereas theology had once provided the wrong answers to questions about the meaning of life, "the right answers now come from evolutionary science."[96]

Such popular accounts of science not only assume the social functions of myth with their attendant moral imperatives, but some also propound their own ersatz eschatologies (scenarios describing the end of the world). These typically amount to one of two not altogether compatible futures — one, a rather dismal dystopia; the other, a kind of brave new world in which religion will have withered away and scientific technology will have alleviated the manifold inconveniences of mortal existence. The more sobering scenario is proffered by the distinguished cosmologist Martin Rees who warns in *Our Final Century: Will the Human Race Survive the Twenty-First Century* (2003) of a variety of potential catastrophes that await the human race. E. O. Wilson speaks in similar terms of an impending environmental crisis. These scenarios, like the epic of evolution, feed into the moral program, for they demand of us repentance and contrition. Other spokesmen for science seem more sanguine about the future. In his anti-religious manifesto, *God Is Not Great* (2007) British-American journalist Christopher Hitchens predicted that "we can consciously look forward to the further evolution of our poor brains, and to stupendous advances in medicine and life extension, derived from work on our elementary stem cells and umbilical-cord blood cells."[97] Fellow traveler Richard Dawkins believes that technological innovation will necessarily be accompanied by moral progress, on account of "a changing moral zeitgeist." Based on an assessment of our present moral superiority—morally speaking "we are way ahead of our counterparts in the Middle Ages, or in the time of Abraham, or even as recently as the 1920s"—we can reason that things will be even better in the future. Thus, "over the longer timescale the progressive trend is unmistakable and it will continue."[98] Subsequent generations will thus achieve their moral superiority not by working hard at the cultivation of the virtues, nor by struggling against their own inclinations in order to adhere to some rule or principle, but simply by being born late in history in a scientifically enlightened age. Most recently, the cognitive scientist and popular science writer Steven Pinker has suggested that violence and warfare are decreasing, and that we live in the most peaceful age of all time. This is attributed partly to "a worldview that we can call *Enlighten-ment Humanism*."[99] Rather than informing moral action, these more utopian scenarios provide reassurance that science will bring in its train a range of material and moral goods.

At one level, these relatively "thin" scientific stories might be thought of as bearing the effete vestiges of the morally robust cosmological traditions of antiquity and the Middle Ages. At another level, they also point to the continuing need for science to have a unifying narrative—some kind of moral or aesthetic vision to promote its relevance to the public. Natural theology had fulfilled these functions for its predecessors, natural history and natural philosophy. But from the late nineteenth century onward, it was a kind of negative version of religion that came for many to take the place of natural theology. The otherwise puzzling antireligious rhetoric of some contemporary scientific popularizers is thus to be understood as a kind of inversion of a long-standing historical relationship that bound the study of nature to deeper moral and religious concerns.

If we were to ask, in light of all this, how modern science got into the truth and goodness business, we could say that this was a legacy of its roots in natural philosophy and natural history, which from ancient times had been concerned with these elevated matters. These latent features of the project of modern science are often obscured by a more recent inheritance from the early modern period in the form of an emphasis on material utility. But the usefulness of science and its claims to authority in the realms of knowledge and morality are intimately connected. In the seventeenth and eighteenth centuries, as we have seen, there was considerable resistance to the shift of focus of natural philosophy away from the contemplation of truth and toward the provision of human comforts. At this time, the practical applications promised by the new science were not self-evident goods that were capable of providing a justification for the novel scientific approaches. As a newly postulated goal of natural philosophy, material usefulness required a further justification that was found in the larger moral and theological framework provided by the Christian story. Utility was aligned with charity, and prospective advances in human welfare with a providential plan of history.

When we consider the contemporary relation between the instrumentality of the sciences and their claims on truth we see the reverse of this pattern of justification. It is the utility of science—the fact it yields practical outcomes and useful technologies—that now provides the basis of its ambitious claims to give us access to a true picture of the world. Science is true, we are repeatedly told, because it works. There is, of course, a subtle slight-of-hand involved in this line of justification, and one that becomes apparent as soon as we consider how many scientific theories and models that have yielded true predictions, practical outcomes, or useful technologies have nonetheless been superseded: Ptolemaic astronomy provided perfectly acceptable predictions of the positions of the heavenly bodies; Newtonian physics still works well enough for much of terrestrial mechanics; James Clerk Maxwell's theory of the ether provided the theoretical basis for the transmission of radio waves. Yet all of these theories have been overtaken—not just incrementally modified and refined, but overtaken in revolutionary ways. The history of science is a graveyard of theories that "worked" but have since been replaced. This conflation of truth and utility is an artifact of the manner in which "science" was aggregated from a range of different activities over the course of the modern period.[100]

The irony in all of this is that it was the formation of the modern notion of religion that made possible the ceding of this territory to science. The capacity of natural philosophy to provide rational support for the propositions that comprised "*the* Christian religion" lent it social legitimacy at a crucial historical moment. The simultaneous investment of utilitarian outcomes with moral and eschatological significance similarly made it possible for justifications of scientific activity to be couched in terms of its practical utility, and these continue to represent the chief grounds for ongoing investment in science.

We feel that even if all possible scientific questions be answered,
the problems of life have still not been touched at all.
—Ludwig Wittgenstein, *Tractatus* 6.52

The rise of the West is, quite simply, the pre-eminent historical
phenomenon of the second half of the second millennium after
Christ. . . . It is only by identifying the true causes of Western
ascendancy that we can hope to estimate with any degree of
accuracy the imminence of our decline and fall.
—Niall Ferguson, *Civilization: The West and the Rest*[1]

Science has not only progressively reduced the competence of
philosophy, but it has also attempted to suppress it altogether and
to replace it by its own claim to universality.
—Nikolai Berdyaev, *Slavery and Freedom*[2]

The Austrian-British philosopher Ludwig Wittgenstein (1889–1951) is well
known for having proposed that certain philosophical problems arise out
of the language in which they are couched, and that these problems disappear once we have untangled the logical knots in our language. On a
partial analogy with this approach, my suggestion in this book has been
that at least some of our contemporary quandaries to do with competing
religious truth claims and conflicts between scientific and religious doc-

trines arise out of the way in which we presently deploy the terms "religion" and "science," and that the history of the way in which these terms came to take on their present meanings will help us see them in a new light. However, I do not think in this case that understanding the history will make conflict between science and religion simply disappear—largely because the respective concepts are so firmly entrenched in our vocabulary that they cannot easily be dispensed with. That said, it is probably worth giving some thought to the ways in which uncritical use of the respective terms is likely to be conducive to a fostering of either unnecessary conflict or the construction of false tensions. Equally important, uncritical deployment of these categories can also lead to the false impressions of congruence. I will return to this issue shortly.

In the meantime I want to point out that Wittgenstein's contributions are relevant to the present task in another way, since the so-called linguistic turn with which he is associated had an important influence on the conduct of intellectual history. This twentieth-century development stressed the way in which our ideas are socially constructed and do not simply map directly onto features of the natural or social worlds. That is not to say that there is no mind-independent reality, but merely that the way in which we conceptualize reality is inevitably influenced by social factors and this is reflected in our language. Two influential approaches to history have embraced this insight. The contextualist approach, advocated by Quentin Skinner and John Pocock, is synchronic, with a tight focus on a specific period, particular historical actors, and a defined geographical setting.[3] Here the idea is that the relevant ideas can only be understood in the context of their linguistic environment. For some time, in the Anglophone world, this has been a dominant mode for pursuing intellectual history. Conceptual history (*Begriffsgeschichte*)—the history of concepts—shares with contextual history a concern with language and its social contexts, but is more ambitious in its historical range. Certainly, it includes a significant synchronic component, since it must establish the meaning of concepts within contemporary semantic fields. But it aims to track changes in these meanings across time, as this book has attempted to do with "science," "religion," and a number of related notions. There are, of course, other ways of doing good history, but what all good historical accounts have in common is the provision of reasonable arguments and sound evidence.[4] I hope that this book conforms to that general standard.

To my mind, the particular advantage of conceptual history is that it shows the relevance of historical analysis to the present while avoiding the attendant dangers of anachronism. The four centuries leading up to the present I have simply characterized as "Western modernity," not without cognizance of the problematic nature of the expression. My point with respect to modernity is that one of the features of its discourse is a particular way of deploying the concepts "science" and "religion." This deployment is distinctive both in relation to our past and to other cultures. Historians can make up their own minds about how my account of the emergence of these two interrelated categories fits in with other large-scale narratives about secular modernity, but here are some brief reflections on that topic.

All Coherence Gone?

In *After Virtue*, the Scottish philosopher Alasdair MacIntyre set out his own diagnosis of Western modernity, focusing on the failure of Enlightenment thinkers to provide grounds for a universal morality. In a compelling illustration of his general thesis, MacIntyre draws on the premise of Walter M. Miller's sci-fi classic, *A Canticle for Leibowitz*. The work is set in a postapocalyptic world in which only the scattered fragments of modern science have survived. Monks of the "Order of Leibowitz" profess the disparate elements of the lost sciences, but without any of the underlying theory that would render them coherent or relate them to one another in a systematic way. MacIntyre's argument is that the post-Enlightenment moral philosophy is in a similar predicament, for despite the continued rehearsal of various moral prescriptions, the theoretical foundation that grounded those prescriptions had been erased with the early modern rejection of Aristotelian teleology. Given this, MacIntyre contends, the modern moral project was doomed to failure from the outset.

MacIntyre did not concern himself much with the emergence of modern science, although it is interesting that his focus on the incommensurability of contemporary moral discourses came after his reading of philosophers of science Thomas Kuhn and Imre Lakatos. It may seem that an implication of MacIntyre's analysis is that science was the beneficiary of this process, since its success was predicated on the abandonment of the teleology that had once rendered moral discourse intelligible. This is partly true, but my suggestion is rather different—namely, that there are significant parallels between the status of modern moral discourse and that of modern science. In other words, the dystopian science of the Brothers of Saint Leibowitz is analogous not only to modern moral philosophy, but to a lesser degree to modern science itself. This is most obviously true from the perspective of the older natural philosophy and is evident in its proponents' critiques of the new science of the seventeenth century. New approaches to the study of nature were said to be no longer properly philosophical and were regarded as falling short of the criteria for true "science." They failed to edify their practitioners and were not connected in a robust way to an integrated view of the cosmos. John Donne lamented that the "new philosophy calls all in doubt" before rendering the verdict: "all coherence gone."[5] It is also worth noting in this context that the marriage of the experimental and mathematical, exemplified in the title of what is perhaps the most famous work of science ever produced— Newton's *Philosophiae Naturalis Principia Mathematica*—was a troubled union. Indeed, Newton is emblematic of the kinds of methodological tensions that beset the new natural philosophy. Rob Iliffe, a leading Newton scholar, has gone so far as to declare that the various elements of Newton's natural philosophy "could not be 'coherent' because the structures in which they appeared were fundamentally incompatible."[6] Admittedly, the centrifugal tendencies of a number of seventeenth- and eighteenth-century sciences of nature were counterbalanced by the unifying force of a natural theology that assumed some of the functions once served by Aristotelian teleology. But when this theological context began to fragment in the nineteenth century, the consequence was "a great empire falling to pieces," to use Whewell's evocative phrase. The various strategies to pull together particular "scientific disciplines" were successful at rhetorical, political, and institutional levels, but, as a number of contemporary philosophers of science have observed, this does not necessarily confer any metaphysical unity on modern science.[7]

Whether one considers modern science, as a category, to be coherent really depends on how one views what took place in the nineteenth century, when the aggregation of a range of activities under the concept "science" took place, attended by the explicit exclusion of others—notably religion and metaphysics. This conglomerated science no longer enjoys the moral and theological justifications that in the early modern period had lent legitimacy to a number of its original constituents. It relies instead on the dual devices of "the scientific method" and "the scientist." And to some extent it also depends upon standing in a particular relation to "religion"—as representing a kind of rational counterpart to an irrational belief system, an alternative source of meaning and value, or a more advanced stage of human development that was destined to replace a more primitive age of religion. Just to be clear, my claim is not that particular sciences or scientific theories are incoherent. Rather I am suggesting that we can reasonably ask whether the natural sciences, taken together, share some common features by virtue of which it becomes possible to speak of some generic thing "science," or whether there is merely a cluster of myths that sustain our present vision of a unified science.

Modern religion's relation to these developments has been threefold. First, the reification of religion is related to the demise of the Aristotelian virtues, as is the reification of *scientia*. That process means that "religion," too, is incoherent, again not in the sense that the activities that it purports to represent are incoherent, but in the sense that "religion" problematically claims to stand for some universal feature of human existence. Second, when experimental natural philosophy and early modern natural history are finding their feet, it is a reified religion that offers them support, and a degree of unity, in the form of a new natural theological project. Third, and somewhat paradoxically, "religion" has now become a contrast case for modern science. Religion is what science is not: a kind of negative image of science, and this contrast has become important for the integrity of the boundaries of science. It follows, to a degree, that the legitimacy of modern science depends on its capacity to compensate for what once was offered by religion, or if not, in demonstrating that we can dispense with it.

The Leibowitz scenario actually applies in a more direct way to the early modern period, since a number of seventeenth-century proponents of experimental science were fully conscious of a preceding catastrophe that had rendered a coherent Aristotelian-style science impossible. But for them the cosmic cataclysm that had long predated the construction of Aristotelian moral and natural philosophy was the primeval Fall. Adam was thought to have possessed an encyclopedic knowledge of the natural world that was now lost as a consequence of original sin. While fragments of this knowledge had been passed on to his posterity, a complete reconstruction was thought to be impossible in the present postlapsarian estate. As the Protestant Reformers pointed out, Aristotle was ignorant of this event and as a consequence his optimistic philosophy had been constructed upon a false foundation. Subsequently, Aquinas had mistakenly assumed that key elements of this philosophical system could be indifferent to this central element of Christian anthropology, and (for his early modern critics) his enthusiasm for Aristotle had unintentionally introduced pagan presuppositions into medieval Christian thought. The early modern scientific project, then, was an attempt at a partial restoration of Adamic knowledge, the difference between the brothers of Leibowitz and the fraternity of seventeenth-century natural philosophers lying in the fact that the latter were aware of the necessarily makeshift nature of their endeavor. Many of those who sought to reconstruct a properly Christianized natural philosophy in the early modern period rejected the sanguine commonsense philosophy of Aristotle, and relinquished his aspirations to a demonstrative science. Natural history and experimental natural philosophy were regarded as fragmented and makeshift enterprises, their fragile status being understood as an inevitable consequence of the cosmic fall that had rendered the operations of nature opaque and compromised human cognitive capacities. As John Locke remarked, in our present condition we have at best "dull and weak" faculties. Accordingly, he concluded, "it appears not, that God intended we should have a perfect, clear, and adequate knowledge."[8] The diffidence of seventeenth-century naturalists was lost in the nineteenth century, when the original reasons for their epistemic modesty were forgotten and the idea of progress became firmly embedded into the West's self-understanding.[9]

Epistemic Grace

Turning to more recent work on the history of science, in the first of a projected series of volumes addressing the theme of *Science and the Shaping of Modernity*, Stephen Gaukroger makes a number of claims that relate to the thesis outlined in this book. First, he also notes that the modern scientific project begins not with a separation of Christianity and natural philosophy, but with the forging of a more intimate relationship between them: "Christianity took over natural philosophy in the seventeenth century, setting its agenda and projecting it forward in a way quite different from that of any other scientific culture, and in the end establishing it as something in part constructed in the image of religion." The success of the new natural philosophy of this period, he observes, was because experimental natural philosophy "could be accommodated to the projects of natural theology."[10] This is consistent with my suggestion that new science is partly shaped by an attempt to re-Christianize natural philosophy, that the legitimacy of the new science depends upon its relationship with Christianity, and that this relationship helps shapes our modern idea "religion." This also relates to the "epistemic turn" that takes place at the beginning of the modern period and to a new tendency to regard religion and religions as characterized by propositional beliefs, which, on a par with beliefs in other spheres, require rational justification. This epistemic turn is a prerequisite for the development of what Gaukroger identifies as "one of the most distinctive features of a scientific culture in modern Europe . . . [namely,] the gradual assimilation of all cognitive values to scientific ones."[11]

To Gaukroger's observations we might add the viewpoints of Taylor and MacIntrye. While his emphasis and approach is different from that of Gaukroger, Charles Taylor has spoken in *A Secular Age* about "new conditions for belief" in the modern age. Specifically, he speaks of "a new shape to the experience which prompts to and is defined by belief; in a new context in which all search and questioning about the moral and the spiritual must proceed." This is the "social imaginary" that underlies our conception of the role of belief.[12] Part of that context is the appearance of a new and distinctive understanding of belief, and the appearance of a

neutral epistemic space, identified with a universal reason. This is further
related to the emergence of modern liberalism, which posits the existence
of a neutral public sphere. Both Taylor and MacIntyre have questioned the
putative neutrality of this public space where no single religious tradition
is favored, suggesting that modern liberalism might be thought of more
along the lines of a competing ideology or religion, asserting its own su-
premacy at the cost of other traditions.[13]

My account suggests a parallel development in the idea of an Archi-
medean epistemic space, in which supposedly neutral rational considerations trump all others. The creation of such a space was partly motivated
by the need to adjudicate between competing truth claims of the "religions," themselves the product of the new conception of religion. It is no
exaggeration to say that this was the chief epistemic concern of the immediate post-Reformation period. Advocates of a religion of reason, insofar as they supported the alliance of natural theology and natural philosophy (and thus a natural theology that was constructed upon "neutral"
rational grounds) were complicit in this development. The possibility
of offering this kind of rational support for religion, where the latter is
understood propositionally, necessitates the creation of a supposedly
neutral space. Initially, that neutral sphere was occupied by natural theology and natural philosophy with their shared physico-theological mission. Over the course of the nineteenth century, however, that territory
was gradually ceded to a coalescing "science." This ultimately resulted in
the assimilation of all cognitive claims to scientific ones. The high-water
mark of this development was the twentieth-century positivist critique of
religious, moral, and aesthetic language. This positivist ethos still lingers,
and the insistence that science sets the standards for what counts as genuine knowledge remains a characteristic feature of the modern Western
epistemological discourse. Arguably, the epistemic imperialism of science
was inherited from the supposedly neutral grounds of eighteenth-century
natural theology from which it emerged.

Max Weber addressed the issue of the universal ambitions of Western
religion and science in the opening lines of *The Protestant Ethic and the
Spirit of Capitalism*: "A product of modern European civilization, studying

any problem of universal history, is bound to ask himself to what combination of circumstances the fact should be attributed that in Western
Civilization, and in Western Civilization only, cultural phenomena have
appeared which (as we like to think) lie in a line of development having
universal significance and value."[14] Weber goes on to identify "science" and
"philosophical and theological wisdom of the most profound sort" as the
phenomena that we most like to imagine as having such universal significance. In other words, the cultural phenomena of the modern West that
have most conspicuously laid claim to universal significance and value are
religion and science.

The acuteness of Weber's observation extends not only to his identification of distinctive characteristics of Western culture, but also to the fact
that despite their claims of universality, these enterprises have a history
and were created out of a particular "combination of circumstances." This
book has sought to describe some of the circumstances that gave rise to
our present understandings of science and religion. It was the construction of the notion "religion" in the early modern period, itself premised
upon a unique understanding of religious identity forged by the early
Christians in the first three centuries of the common era, that provided
the prototypical model of a belief system for which is claimed universal
and transcultural significance. The new conception gave rise to a new kind
of problem—that of the conflicting truth claims of other "religions," constructed in the image of Christianity. This was followed by the construction of the neutral epistemic space in which Christianity was "impartially"
judged to be *the* true religion. In the nineteenth century, the assumption
of universal applicability came to be built into the fabric of the newly constructed "science." At this juncture in history the West's sense of the source
of its superiority "shifted seamlessly" from its religion to its science.[15] The
new conception of science changed the locus of conflict from competing religious truth systems to the competing truth claims of science and
religion. The birth of the conflict model, and its projection back into history, were thus consequences of this transfer of epistemic authority from
religion to science. One of the reasons that our science makes universal
claims, then, is that it borrows from "the Christian religion" its notions

of universal applicability. The modern idea of religion made it possible for Christianity to claim to be the one true religion. Modern science now claims an analogous universal applicability.

It goes without saying that the contingency of the historical circumstances that gave rise to these features of Western culture is not always easy to square with the universal claims that they make. In the case of Christianity this is a "known problem." The "scandal of particularity" for Christianity concerns how divine workings in a specific historical time and place might be given a universal and timeless significance. The same difficulty was articulated in another guise by Gotthold Lessing (1729–81), in the celebrated dictum that "accidental truths of history can never become the proof of necessary truths of reason."[16] Ferdinand Christian Baur (1793–1860) concerned himself with a similar set of problems, demonstrating that the emergence of Christian orthodoxy in the first centuries of the common era was the product of conflict and negotiation between different parties over long periods of time. Baur sought to bring church history into the realm of naturalistic explanation, and to some degree withheld judgment about which competing parties' sides were theologically "correct." The basic assumption informing the work of Baur and the Tübingen school was that the historical triumph of core elements of Christian dogma could not be explained simply by claiming that they are true. Only in the latter half of the twentieth century did historians begin to ask the same questions about the development of science.[17]

It may be observed in this connection that Christianity and science have resisted the telling of their histories. The very first ecclesiastical historian, Eusebius of Caesarea, proceeded on the assumption that heresies had a history, but not Christianity. There was an ideal and eternal form of Christianity that had existed since the beginning of time, albeit shrouded within a variety of cultural forms and practices. We see this also in Augustine's insistence that "true religion" had always been present from the time of creation, albeit in a variety of implicit forms.[18] Divine providence, working in accordance with its own logic, ensured the preservation of this true religion, while deviations from it were to be accounted for by other factors. So-called internalist histories of science followed a very similar pattern, assuming that the progress of science entails the unfolding of eternal truths, and that the only genuine contingency in the history of science is to do with whether the requisite genius is present at any particular historical time to discover them. In this style of history, historical actors are rational insofar as their discoveries anticipate what will become scientific orthodoxy, and irrational if they deviate from the lineage of modern science. Historical and sociological explanations are needed only for the latter, for the former are understood as ongoing developments of a logical argument. As philosopher Larry Laudan has expressed it: "When a thinker does what is rational to do, we need enquire no further into the causes of his action, whereas when he does what is in fact irrational—even if he believes it to be rational—we require some further explanation."[19] Historians of science have long since abandoned internalist and "long march of progress" accounts of science. Yet these still inform popular representations of the history of science, and underpin claims for the epistemic superiority of its methods.

While there are interesting similarities between the histories of religion and of science, there is also one significant difference. The appeal to divine providence as guarantor of the eventual emergence of truth, while consistent with the theistic assumptions of those who hold to the truth of religion, seems less appropriate in the history of science, particularly for those committed to scientific naturalism. The more absolutist claims about scientific knowledge, not to put too fine a point on it, have a whiff of theological dogmatism about them. As David Bloor has pointed out, internalist history of science assumed the form of the history of apostolic truth.[20] If the rise to cultural authority of science in the modern period contributed to the hypostasization of religion, the kind of authority that science assumed was borrowed from the magisterial cultural authority that had once belonged to religion. Modern science has thus inherited the *form* of the universal claims of Christianity without the supernaturalist foundation that would render them internally coherent. Science, thus understood, enjoys none of the epistemic comforts afforded to the supernaturalist.[21]

The Conflict Myth?

I suggested at the outset that although science-religion conflict can be
understood as the outcome of the categories themselves, the option for
dispensing with them is not realistic. Yet there is certainly room for a more
nuanced and critical deployment of the terms. Some (admittedly extreme)
examples might help make this general point. A recent book that seeks
to mount a case "against religion" understands its target to include the
following: Mother Teresa, voodoo, the pope, fear-ridden peasants of antiquity, Muslim suicide bombers, animists, arid monotheism, the archbishop of Canterbury, séances, Thomas Aquinas, an evangelical huckster
dressed in a Little Lord Fauntleroy suit, Muhammad, the tawdry myths of
Bethlehem, the vapid and annoying holiday known as Hanukkah, Mor-
mons, hysterical Jewish congregations, the sordid theology of Pascal,
Martin Luther King, rednecks, cobbled-together ancient Jewish books,
WWII–era Japanese emperor worship, and male circumcision.[22] The list is
not exhaustive. From the same genre, here is a definition of science: "Science, in the broadest sense, includes all reasonable claims to knowledge
about ourselves and the world."[23] Not surprisingly, the authors of these
definitions are firmly committed to the view that science and religion are
cultural constants, and that conflict between them is inevitable.

These crude characterizations come out of a particular polemical context, which it is not my intention to engage directly. One way of chal-
lenging these claims, however, would be to question what are obviously
contestable definitions. This steers the discussion toward issues of how
religion and science should be properly defined. My own suggestion has
been that the issue is not to do with how to populate these two categories,
but with the fact that the discussion is framed by these categories to begin
with. As I hope is now apparent, science and religion are not natural
kinds; they are neither universal propensities of human beings nor necessary features of human societies. Rather they are ways of conceptualizing
certain human activities—ways that are peculiar to modern Western culture, and which have arisen as a consequence of unique historical circumstances. So while this historical analysis may not make science-religion
conflict go away, it should be clear why it has emerged at this particular

time and place. Returning to the issue of contestable definitions, the fact that science and religion are not natural kinds means that there are no firm criteria for adjudicating what should or should not be included in the concepts. As the examples above illustrate, this makes it possible to generate perceived conflicts simply by filling out these incorrigible categories in a particular way.

This is relevant to broader discussions of whether religion is a good thing and whether we can offer naturalistic explanations of religion in order to explain it, or, to explain it away. A further implication of the history outlined above is that religion is not a very good candidate for an explanatory variable *or* as something to be explained. On the latter point, recent years have seen a number of attempts to offer naturalistic accounts of religion that purport to explain why it arose and why it persists. One early and influential account was provided by the Scottish philosopher David Hume. At the very outset of his *Natural History of Religion* (1757), Hume observes that with regard to religion "there are two questions in particular which challenge our principal attention, to wit, that concerning its foundation in reason, and that concerning its origin in human nature."[24] Hume helpfully illustrates two features of the emerging object "religion": first, that it is understood as akin to a "natural kind" and that, as such, naturalistic explanations can be offered to account for it; second, that religion is susceptible to rational analysis and that it can be established whether or not the propositions that constitute it are grounded in reason. Religion can now provide the subject matter of what we would now call the "science of religion" or "history of religions," on the one hand, and "the philosophy of religion," on the other. Hume here assumes that there is a thing, religion, that is primarily constituted by beliefs, and about which we can ask two questions: are these beliefs warranted, and what kind of naturalistic explanation can we offer for their having arisen?

While these two questions have at times been conflated by Hume's successors, Hume was himself astute enough to realize that they are quite distinct. To offer a naturalistic account of religious beliefs is not to explain them away. The warrant for believing something is independent of the mechanisms through which we came to acquire the beliefs. Moving into the present, the kinds of enterprises that position themselves as successors to Hume's pioneering endeavors are works in the "religion as a natural phenomenon" genre.[25] Here again, the aim is to offer naturalistic explanations for the ubiquity of religion. As I hope is now obvious, while it might be possible to offer explanations of, say, the emergence of the cognitive substrate that is a precondition for the appearance of some elements of what we include in the category religion, it will not be possible, in principle, to explain religion, because religion is not a single thing, but rather a cultural construct of the modern West. The only legitimate explanation of religion is the kind of explanation that historians offer—that is, a history of its appearance as a category. To advert to the analogy set out in chapter 1, we cannot in principle offer a naturalistic account of religion any more than we can offer a single explanation of the geological and chemical processes that give rise to the natural formation of jade—since "jade" is in reality two different things. Religion, similarly, is a number of different things that we have decided to lump together. And this precludes there being any single naturalistic explanation of religion's origins.[26]

It follows, by the same token, that religion cannot serve as an explanation for anything, either. To take a single example—the association of religion and violence—it can be said that while it might be possible to establish connections between violence and elements that have been included in the *category* religion, there remains the question of whether these elements should be so categorized in the first place.[27] This applies to any number of instances—conflict in Northern Ireland, the early modern "wars of religion," the crusades, and the menace of "Islamic terrorism" to mention just a few. Hypotheses that link religion and violence are unhelpful not simply because the categories are confused, but also because they prevent us from understanding the true complexion of causes of those ills for which a constructed "religion" is the convenient scapegoat. This same difficulty afflicts the Samuel Huntington "clash of civilizations" hypothesis, which similarly relies upon overly simplistic constructions.[28] While we persist with these false categories, we will be prevented from discerning the true causes of the difficulties that presently beset us.

Finally, while it is clear that there cannot have been a perennial warfare between science and religion, the conflict myth continues to serve the role for which it was originally fashioned in the late nineteenth century, of establishing and maintaining boundaries of the modern conception "science." The present battle lines, ironically, take their shape from a previous synthesis of natural theology and the natural sciences. Popular critiques of religion that enlist science in their cause might thus be understood as representing a persistence of the classic form of nineteenth-century popular science that used natural theology as a medium to provide narrative coherence and real-life relevance to broad general audiences. In the present case, however, atheism assumes the role once occupied by natural theology. This accounts for the conjunction of atheism and science that one encounters in at least some recent works of popular science. While the ostensible focus in high profile science-religion disputes is factual claims about the natural world, such debates are often proxies for more deep-seated ideological or, in its broadest sense, "theological" battles. For their part, what religiously motivated antievolutionists fear is not the "science" as such, but the secularist package of values concealed in what they perceive to be the Trojan horse of evolutionary theory. Perhaps these skirmishes should be thought less in terms of conflict between science and religion, and more as theological controversies waged by means of science. Such conflicts are, again, irresolvable, not because there is any inherent incompatibility of science and religion, but because the underlying value systems—which are "natural theologies" of a kind—are ultimately irreconcilable. Given this situation, it is interesting to reflect upon whether the contemporary amalgam "science" needs to be sustained by an underlying value system in the same way that early modern science was sustained by its framework of natural theology. Naturalism or atheism may be able to perform this role. But one reason for doubting this relates to a general intellectual difficulty of atheism as an ideology (as opposed to a personal conviction), which is its reliance on the more systematic belief structures that it seeks to negate. In other words ideological atheism is parasitic on the modern construct "religion," understood as constituted by propositional beliefs that require rational justification.

It is worth observing, finally, that proponents of conflict are not the only ones promoting the viability of "science" and "religion." Advocates of positive relations between science and religion, who argue that science supports religious belief, also act to reinforce the modern boundaries of "science" and "religion." Much like the "secular theologians" of the early modern period, their urging of a consonance between science and religion has the potential to reinforce the very conditions that make conflict possible. Advocates of constructive dialogue are thus unknowingly complicit in the perpetuation of conflict. Often, they concede the cultural authority of the sciences, the propositional nature of religion, and the idea of a neutral, rational space in which dialogue can take place. As we have seen, each of these developments is relatively recent. But the history of their emergence, along with the past from which they came, offers some intriguing intimations of how things might have been, and might yet be, rather different.

[ACKNOWLEDGMENTS]

Over the long gestation of ideas in this book I have accumulated numerous debts. Just over fourteen years ago the Center of Theological Inquiry at Princeton offered me a year-long Witherspoon Fellowship in science and religion. During that fellowship I first began to develop the ideas that form the basis of this book, before other projects intervened. I am grateful to the Witherspoon family and the CTI and its 2001 fellows for their support during this gestational period. The impetus for resuming the project was my election to the Gifford Lectureship for 2011, and I am indebted to the Gifford committee for the honor of the lectureship and opportunities that it provided. In 2010–11, the UK's Arts and Humanities Research Council also awarded me a research fellowship that afforded me time to write and research the lectures that form the basis of the book. The University of Oxford was generous in granting me leave during this period to take up the fellowship, and I am grateful to my colleagues in the Faculties of Theology and History, and Harris Manchester College, for being willing to assume some of my duties during my absence. Conversations with numerous colleagues and friends also proved helpful. I am particularly grateful to Dan Garber, who read the manuscript in its entirety and offered invaluable criticism. In no particular order, thanks are also owing to Ian Hunter, Stephen Gaukroger, Philip Almond, Rick Strelan, Andrew Moore, Allen Verhey, Richard Yeo, Ron Numbers, Jon Roberts, John Brooke,

Rhodri Lewis, Will Poole, Pietro Corsi, Guy Stroumsa, and Matt Walhout. The usual caveat applies—that the assistance rendered does not imply acquiescence in the views propounded in the book. Versions of the lectures were discussed in reading groups at Oxford and Duke, from which I received invaluable feedback. A special thanks to Ignacio Silva and Joseph Wolyniak for organizing those groups. I am also grateful to the audience at Edinburgh and interlocutors among the faculty there, in particular, David Fergusson, Stewart Brown, Larry Hurtado, Oliver O'Donovan, Zenon Bankowski, Michael Fuller, Wilson Poon, and the vice chancellor, Professor Sir Timothy O'Shea. These individuals were erudite conversation partners and warm and congenial hosts. The readers for University of Chicago Press offered very helpful remarks, and the acquisitions editor for history of science, Karen Merikangas Darling, has been a great source of encouragement. Finally, I must thank my family (Carol, Grace, and Thomas) for their forbearance over the years that I have been writing this book, and in particular for their willingness to accompany me to far-flung parts of the world over the course of my academic career. I owe a special debt to Carol, who has gone well beyond the call of duty in supporting my writing, including reading through this entire manuscript. This book is dedicated to Thomas.

[ABBREVIATIONS]

ACW	*Ancient Christian Writers* (New York: Paulist Press)
ANF	*Ante-Nicene Fathers* (Edinburgh: T. & T. Clark, 1989)
Aquinas, SCG	Thomas Aquinas, *Summa contra gentiles*, 4 vols. (Notre Dame: University of Notre Dame Press, 1975)
Aquinas, ST	Thomas Aquinas, *Summa theologiae*, Blackfriars ed. (London, 1964–76)
Augustine, Works	*The Works of Saint Augustine*, 20 vols., ed. John Rotelle (New York, 1991–)
Bacon, Works	*The Works of Francis Bacon*, 14 vols., ed. James Spedding, Robert Ellis, and Douglas Heath (London: Longman, 1857–74).
CCSL	*Corpus Christianorum Series Latina*
CSEL	*Corpus Scriptorum Ecclesiasticorum Latinorum*
Descartes, CSM	*The Philosophical Writings of Descartes*, 2 vols., trans. J. Cottingham, R. Stoothoff, and D. Murdoch (Cambridge: Cambridge University Press, 1985)
Descartes, CSMK	*The Philosophical Writings of Descartes* (vol. 3, *The Correspondence*), trans. J. Cottingham, R. Stoothoff, D. Murdoch, and Anthony Kenny (Cambridge: Cambridge University Press, 1991)

FC   *Fathers of the Church* (Washington, DC: Catholic
                            University of America Press, 1947–)

LCL   *Loeb Classical Library*

NPNF I   *Nicene and Post-Nicene Fathers, First Series*, ed. Philip
                            Schaff (New York, 1886–)

NPNF II   *Nicene and Post-Nicene Fathers, Second Series*, ed. Philip
                            Schaff (New York, 1886–)

PG   *Patrologia cursus completus, series Graeca*, ed. J.-P. Migne
                            (Paris, 1857–1912)

PL   *Patrologia cursus completus, series Latina*, ed. J.-P. Migne
                            (Paris, 1857–1912)

[NOTES]

A Note on the Graphs

1. http://quod.lib.umich.edu/e/eebogroup/, accessed 18 July 2012.

2. For a fuller explanation, see http://books.google.com/ngrams/info, accessed 18 July 2012; and Jean-Baptiste Michel, Yuan Kui Shen, Aviva Presser Aiden, Adrian Veres, Matthew K. Gray, The Google Books Team, Joseph P. Pickett, Dale Hoiberg, Dan Clancy, Peter Norvig, Jon Orwant, Steven Pinker, Martin A. Nowak, and Erez Lieberman Aiden, "Quantitative Analysis of Culture Using Millions of Digitized Books," *Science* 331, no. 6014 (14 January 2011): 176–82.

Chapter One

1. Seneca, *Natural Questions* 1, Pref. 9.

2. Clement of Alexandria, *Stromata* 7.4 (ANF vol. 2, p. 529).

3. On the distinction between map and territory, see Alfred Korzybski's *Science and Sanity: An Introduction to Non-Aristotelian Systems and General Semantics* (Lancaster, PA: International Non-Aristotelian Library, 1941); Jonathan Z. Smith, *Map Is Not Territory: Studies in the History of Religions* (Leiden: Brill, 1978).

4. See J. LaPorte, *Natural Kinds and Conceptual Change* (Cambridge: Cambridge University Press, 2004), pp. 94–100. On Israel and ideals of nationhood, see Shlomo Sand, *The Invention of the Land of Israel: From Holy Land to Homeland* (London: Verso, 2012).

5. Plato, *Phaedrus* 265e.

6. Thomas Aquinas, ST 2a2ae, 81, 5–6. For the full range of meanings of *religio* in Aquinas, see Roy J. Deferrari et al., *A Lexicon of St. Thomas Aquinas* (Baltimore: John Lucas, 1948), p. 960.

7. A second sense of *religio* in the Middle Ages refers to the monastic life. This is the first meaning of the English "religion" offered by the *Oxford English Dictionary*: "A state

of life bound by religious vows; the condition of belonging to a religious order." OED online, s.v. "religion," 1. http://www.oed.com/view/Entry/161944?redirectedFrom=religions#eid, accessed 12 June 2012. On these senses of the word, see John Bossy, "Some Elementary Forms of Durkheim," *Past and Present* 95 (1982): 3–18; Peter Biller, "Words and the Medieval Notion of 'Religion,'" *Journal of Ecclesiastical History* 36 (1985): 351–69.

8. Aquinas, ST 2a2ae, 92, 1. For Cicero, *superstitio* was directed toward the appropriate object—God or the gods—but was inappropriate on account of being excessive. *De natura deorum* I, 42, 117; II, 28, 72; cf. Isidore, *Etymologiae* VIII.iii.6. Theologians from the eleventh to the thirteenth centuries typically understood *religio* in this way, as a virtue that was subsidiary to justice. See O. Lottin, *Psychologie et morale au XII et XIII siècles*, 7 vols. (Louvain-Gembloux: Duculot, 1942–54), vol. 3/2, pp. 313–26. Aquinas's discussion of *religio* gained considerable influence through its incorporation into Vincent of Beauvais's great medieval encyclopedia, the *Speculum Maiius*. See Vincent of Beauvais, *Speculum morale* 1, 3.61, in *Speculum quadruplex naturale, doctrinale, morale, historiale*, 4 vols., Douai 1624, vol. 4, pp. 357–60.

9. "*Veram religionem veri dei.*" Tertullian, *Apologeticus* 24.1 (PL 1, col. 476; LCL 250, p. 130), cf. 35.1, *De Spectaculis* 1. Here I differ from the reading of Maurice Sachot for whom Tertullian introduces a new conception of *religio* and endorses Christianity as "the true religion" in a more modern sense. *Quand le christianisme a changé le monde* (Paris: Odile Jacob, 2007), pp. 108–9. Cf. Sachot, "«Religio/Superstitio», Historique d'une subversion et d'un retournement," *Revue d'histoire des religions* 208 (1991): 355–94. Competing with Tertullian as the first Christian writer to use the expression "true religion" is Minucius Felix, *Octavius* 1.5, 38.7 (LCL 250, pp. 314, 434), who uses the expression in the same sense. The influence of one text on the other is likely, but the exact date of composition of *Octavius* is uncertain.

10. Lactantius, *Divine Institutes* 1.1, 2.1 (ANF vol. 7, pp. 9, 40); italics mine.

11. "*Religio munda et inmaculata apud Deum et Patrem haec est visitare pupillos et viduas in tribulatione eorum inmaculatum se custodire ab hoc saeculo*" (Vulgate).

12. This contrast is established in the previous verse: "If any one thinks he is religious but does not bridle his tongue but deceives his heart, this man's religion is vain" (1.26).

13. MS Bodley 820, fol. 72r., Bodleian Library, Oxford. Cited in Biller, "Medieval Notions of Religion," 357.

14. "*In timore igitur et amore Dei, et in observantia mandatorum ejus, consistit vera religio.*" *In epistolas et evangelia dominicalia homiliae* 62, PL 155, col. 1894B.

15. "*In quo multipliciter et copiosissime disputatur, unum verum Deum . . . religione vera colendum.*" Augustine, *Retractionum libri duo* I.13.1. PL 32, col. 602. English translation in *Augustine: Earlier Writings*, ed. John H. S. Burleigh (London: SCM, 1953), p. 218.

16. Augustine, *De vera religione* IV (*Vera religio in sola Ecclesia catholica*). PL 34, col. 0127.

17. "*Nam res ipsa quae nunc christiana religio nuncupatur, erat apud antiquos, nec defuit ab initio generis humani, quousque ipse Christus veniret in carne, unde vera religio quae jam*

erat coepit appellari christiana." Retractionum I.13.3, PL 32, col. 603. "For what is now called the Christian religion existed of old and was never absent from the beginning of the human race until Christ came in the flesh. Then true religion which already existed began to be called Christian." *Augustine: Earlier Writings,* p. 218.

18. Augustine, Letter 102, Augustine to Deogratias 19, *Works,* vol. II/2, p. 30.

19. Ibid., pp. 25, 26.

20. Augustine, Sermon 71, 32, *Works,* vol. III/3, p. 266. Also *Expositions of the Psalms,* Psalm 47, 8, 13, 14, *Works,* vol. III/16, pp. 343, 347, 348. Augustine elsewhere identifies *religio* with piety [*pietas*], devotion [*deuotio*], and true testimony [*uerum testimonium*]. Sermon 335a, 1, *Works,* vol. III/9, p. 211.

21. Marsilio Ficino, *Opera,* 2 tom. (Basel, 1574), I, 6. For brief discussions of Ficino on religion, see Wilfred Cantwell Smith, *The Meaning and End of Religion* [1962] (London: SPCK, 1978), pp. 32–34; Peter Harrison, *"Religion" and the Religions in the English Enlightenment* (Cambridge: Cambridge University Press, 1990), pp. 12–13. Similar sentiments had been expressed previously. Justin Martyr thus wrote that those who lived before Christ according to "reason" or "the word" (*meta logou*) were Christians. I *Apology* 46. Cf. Hugh of Saint Victor, who declared that in all times, from the beginning to the end of the world, there could be found individuals expressing faith in Christ. "*Credimus enim nullum tempus esse ab initio mundi usque ad finem saeculi, in quo non inveniantur fideles Christi.*" *De arca noe mystica,* cap. 3, PL 176, col. 0685B; cf. "*Unde patet quod ab initio et si non nomine, re tamen Christiani fuerunt.*" *De sacramente christiane fidei* VIII, 11; PL 176, col. 0312D.

22. Ulrich Zwingli, *The Latin Works of Huldreich Zwingli,* ed. Clarence N. Heller, 3 vols. (Philadelphia: Heidelberg Press, 1929), vol. 3, p. 56. See Smith, *Meaning and End of Religion,* pp. 35–37.

23. John Calvin, *The Institution of Christian Religion* (London, 1561).

24. Calvin, *Institutes* (1536), trans. Henry Beveridge, 2 vols., Prefatory Address (London: James Clarke, 1962), vol. 1, p. 3.

25. Smith, *Meaning and End of Religion,* p. 51. See also Harrison, *"Religion" and the Religions;* Michel Despland, *La religion en occident: évolution des idées et du vécu.* (Montreal: Fides, 1979); Nicholas Lash, *The Beginning and End of "Religion"* (Cambridge: Cambridge University Press, 1996); Russell McCutcheon, "The Category 'Religion' in Recent Publications: A Critical Survey," *Numen* 42 (1995): 285–30; Ernst Feil, *Religio: Die Geschichte eines neuzeitlichen Grundbegriffs vom Frühchristentum bis zur Reformation* (Göttingen: Vandenhoeck and Ruprecht, 1986); Guy Stroumsa, *A New Science: The Discovery of Religion in the Age of Reason* (Cambridge, MA: Harvard University Press, 2010).

26. Aquinas, ST 1a2ae, 49, 1; 1a2ae, 50, 3; 1a2ae, 52, 2; 1a2ae, 53, 1. Cf. *On the Division and Methods of the Sciences, Questions V–VI of the Commentary on Boethius' De Trinitate* Q. 5, art. 1.1. Thomas's teacher, Albertus Magnus, had similarly pointed out that "sciences are a characteristic [*habitus*] of the intellect." Albert the Great, *Quaestiones super de animalibus* Bk. 1, Q. 1 (FC Medieval Continuation, vol. 9, p. 14). Both follow Aristotle: "Science [ἐπιστήμη] is a capacity to demonstrate." *Nicomachean Ethics* 1139b31–35. For Aquinas on *scientia,* see also Eleonore Stump, "Aquinas on the Foundations of Knowl-

edge," in *Aristotle and His Medieval Interpreters*, ed. Richard Bosley and Martin Tweedale, *Canadian Journal of Philosophy*, suppl., 17 (1991): 125–58; Scott MacDonald, "Theory
of Knowledge," in *The Cambridge Companion to Aquinas*, ed. Norman Kretzmann and
Eleonore Stump (Cambridge: Cambridge University Press, 1993), pp. 188 f., n. 13, and
passim; Aristotle, *Categories* VIII. See also Lewis and Short, *A Latin Dictionary*, s.v.
"scientia": "*a knowing or being skilled in* any thing, *knowledge, science, skill, expertness*"
(p. 1642).

27. This is Aristotle's position. *Nicomachean Ethics* 1103a14–25; *Eudemian Ethics*
1220a; *Magna Moralia* 1184b–1185a.

28. Aristotle, *Metaphysics* 994a; *Nicomachean Ethics* 1094a. Cf. Aquinas, SCG III/1,
16.

29. Aristotle, *Metaphysics* 980a 22.

30. Aquinas, ST 1a2ae, 55, 1; 56, 3; 57, 1. Cf. Aristotle, *Nicomachean Ethics* 1139b
14–18. In his commentary on Aristotle's *Posterior Analytics*, Aquinas notes that there
are five dispositions that relate to what is true: art, *scientia*, wisdom, prudence, and
understanding. ("*Ad cuius evidentiam sciendum est quod Aristoteles in VI Ethic. ponit
quinque quae se habent semper ad verum, scilicet artem, scientiam, sapientiam, prudentiam
et intellectum.*") *Super Post. An.* I, lectio 44, n. 11. Prudence and art related to the practical intellect; science, understanding, and wisdom to the speculative intellect.

31. Aquinas, ST 1a2ae, 57, 2. More specifically, *scientia* was understood to be "cogni-
tion acquired through demonstration." *Super Post. An.*, II, lectio 1.

32. Aquinas, ST 1a2ae, 52, 2. Cf. 1a2ae, 54, 4; Aristotle, *Metaphysics* 994a 30.

33. John Securis, *Detection and querimonie of the daily enormities and abuses comitted
in physick* (London, 1566), quoted in Deborah Harkness, *The Jewell House* (New Haven:
Yale University Press, 2007), p. xv.

34. "Scientia . . . Knowledge, learning, skill, cunning, properly the act of him that
knoweth, secondly the state of the thing known, . . . thirdly an habit of knowledge got
by demonstration, Arist 7. Ethic. 3. fourthly any habit of the understanding." Thomas
Holyoake, *A Large Dictionary in Three Parts* (London, 1676), s.v. "scientia" (unpaginated). Cf. Franco Burgersdijck, *Monitio Logica* (London, 1697), p. 99. Charles Lohr
writes that "the habit of science was thought of as being acquired by an act of demonstration and increasing as more and more conclusions were drawn from principles."
"Metaphysics," *Cambridge History of Renaissance Philosophy*, ed. Charles B. Schmitt
and Quentin Skinner (Cambridge: Cambridge University Press, 1988), p. 632. See also
Heikki Mikkeli, *An Aristotelian Response to Renaissance Humanism: Jacopo Zabarella on
the Nature of Arts and Sciences* (Helsinki: SHS, 1992), pp. 27–29.

35. René Descartes, "Letter to Hogelande," 8 February 1640, in CSMK, p. 144, AT
vol. 3, p. 722.

36. The seven gifts are mentioned in Isaiah 11:2 (Septuagint): wisdom, understanding, counsel, strength, knowledge, godliness, and fear. The fruits of the spirit are listed
in Galations 5:22–23: love, joy, peace, patience, gentleness, goodness, faith. Aquinas,
ST 1a2ae, 68, 1; 70, 1; 2a2ae, 45, 4. On the significance of these categories for Aquinas,

see Andrew Pinsent, *The Second Person Perspective in Aquinas's Ethics* (London: Routledge, 2012).

37. Eleonore Stump, *Aquinas* (London: Routledge, 2003), pp. 360, 352. For the influence of these ideas in the early modern period, see John Cottingham, "Descartes as Sage: Spiritual *askesis* in Cartesian Philosophy," and Peter Harrison, "The Natural Philosopher and the Virtues," both in *The Philosopher in Early Modern Europe: The Nature of a Contested Identity*, ed. Conal Condren, Stephen Gaukroger, and Ian Hunter (Cambridge: Cambridge University Press, 2006), pp. 182–201, 202–28.

38. As noted above (n. 7), in the Middle Ages, there was a specific sense of plural "religions," understood as referring to monastic orders or religious houses.

39. Aristotle, *Posterior Analytics* 78a22–79a33. On the meanings of *scientia* in the Middle Ages, see Steven Livesey, "*Scientia*," in *Medieval Science, Technology, and Medicine: An Encyclopedia*, ed. Thomas Glick, Steven J. Livesey, and Faith Wallis (New York: Routledge, 2005), pp. 455–58; Michael H. Shank, "Natural Knowledge in the Latin Middle Ages," in *Wrestling with Nature*, ed. Peter Harrison, Ronald Numbers, and Michael Shank (Chicago: University of Chicago Press, 2011), pp. 83–115.

40. These were divided into the verbal sciences (*scientiae sermocinales*) and real sciences (*scientiae reales*), with the *trivium* (grammar, logic, rhetoric) comprising the verbal, and the *quadrivium* (arithmetic, astronomy, music, geometry) the real. Shank, "Natural Knowledge," pp. 90 f.

41. Aquinas, *Division and Methods of the Sciences* Q. 5, art. 1. Aquinas observes that there are two possible threefold divisions of the sciences—one in terms of the three intellectual virtues, and another in terms of the nature of the discrete objects of the sciences. The respective objects of the three speculative sciences will be further discussed in ch. 2. Cf. Aquinas, ST 1a, 1, 1.

42. Kilwardby, *De ortu scientiarum* 194–95, cited in Oga Weijers, "L'appellation des disciplines dans les classifications des sciences aux XIIᵉ et XIIIᵉ siècles," *Archivum Latinitatis Mediae Aevi* 46/47 (1988): 42–43.

43. See, e.g., Ulpian Fulwell, *The First Parte, of the eighth liberall science: entituled, Ars adulandi* (London, 1597). For a typical list of the sciences in seventeenth-century England, see D. Abercrombie, *Academia Scientiarum; or, The Academy of the Sciences, being a Short and Easy Introduction to the Knowledge of Liberal Arts and Sciences* (London, 1687).

44. The expertise of a science tutor was thus described in these terms: "not for cutting up cats, but what science meant then, Ethics, Butler and such like." *OED* online, s.v. "science," †c. http://www.oed.com/view/Entry/172672?redirectedFrom=science#eid, accessed 12 July 2012.

45. *Encyclopaedia Britannica* (Edinburgh, 1771), 3 vols., s.v. "Science" (III, 570a). Cf. Chambers, *Cyclopaedia*, I, vii. On early modern conceptions, see Tom Sorell, G. A. J. Rogers, and Jill Kraye, eds., *"Scientia" in Early Modern Philosophy* (Dordrecht: Springer, 2010).

46. Aquinas did address the question of the relation between *scientia* and *fides*,

concluding that since science is concerned with self-evident principles, and faith with unseen ones, "faith and science are not about the same things." ST 2a2ae, 1, 5.

47. Practitioners of *Begriffsgeschichte* (the history of concepts) thus seek to distinguish between "onomasiology," in which different words refer to similar concepts, and "semasiology" in which one word might refer to a number of different concepts. Mark Bevir, *"Begriffsgeschichte," History and Theory* 39 (2000): 273–84.

48. Something very much like this was the first article of the first question of the *Summa theologiae*, viz., "Whether, besides philosophy, any further doctrine is required." ST 1a, 1, 1. The question was also raised by the 1277 condemnation of 219 philosophical and theological theses. For the text of the condemnation, see David Piché, ed., *La condemnation parisienne de 1277. Texte latin, traduction, introduction et commentaire* (Paris: Vrin, 1999). Among many commentaries, see, e.g., Jan A. Aertsen, Kent Emery Jr., and Andreas Speer, eds., *Nach der Verurteilung von 1277. Philosophie und Theologie an der Universität von Paris im letzten Viertel des 13. Jahrhunderts. Studien und Texte* (Berlin: de Gruyter, 2001); Edward Grant, "The Condemnation of 1277, God's Absolute Power, and Physical Thought in the Late Middle Ages," *Viator* 10 (1979): 211–44; John F. Wippel, "The Condemnations of 1270 and 1277 at Paris," *Journal of Medieval and Renaissance Studies* 7 (1977): 169–201.

49. Plato, *Republic* 379a. Aristotle, *Metaphysics* 1025b. Werner Jaeger claims that Plato coined the term "theology," although it is possible that it was already in use at the time. *The Theology of the Early Greek Philosophers* (Oxford: Clarendon, 1948), p. 4. Cf. Gregory Vlastos, "Theology and Philosophy in Early Greek Thought," *The Presocratics*, ed. Daniel W. Graham (Princeton: Princeton University Press, 1993), p. 8, n. 22.

50. "There are two kinds of theology, or divine science . . . the kind of theology pursued by the philosophers . . . [and] the theology taught in the Sacred Scripture." Aquinas, *Division and Methods of the Sciences* Q. 5, art. 4, 8 (Maurer ed., p. 52). In ST 1a, 1, 1 Aquinas makes it clear that the subject matter of the *Summa theologiae* is "sacred doctrine," which is his preferred term. For "theology" used as a derogatory term by Christians, see Stephen Brown, "Key Terms in Medieval Theological Vocabulary," in *Méthodes et instruments du travail intellectuel au moyen âge*, ed. Olga Weijers (Turnhout: Brepols, 1990), 82–97.

51. Aquinas, ST 1a, 1, 3. On the idea of theology as a habit, see John I. Jenkins, *Knowledge and Faith in Aquinas* (Cambridge, 1997), p. 5; Aidan Nichols, *The Shape of Catholic Theology: An Introduction to Its Sources* (Edinburgh: T. & T. Clark, 1991), p. 13, n. 1; Gerald van Ackeran, *Sacra Doctrina* (Rome: Catholic Book Agency, 1952), pp. 51 f.

52. Bonaventure, *Commentaria in Quatuor Libros Sententiarum, Opera Omnia S. Bonaventurae* (Quaracchi: Ad Claras Aquas, 1882–1902), I:13.

53. John Duns Scotus, *Ordinatio* prol. 5.1–2, nn. 314, 332, in *Opera Omnia*, ed. C. Balíc et al. (Vatican City: Typis Polyglottis Vaticanis, 1950–), 1:207–8, 217, quoted in Richard Cross, *Duns Scotus* (Oxford, 1999), p. 9. Cf. Roger Bacon, "all speculative philosophy has moral philosophy for its end and aim. . . . For this reason speculative philosophy is able to prepare the principles of moral philosophy. . . . . With Christian students of philoso-

phy moral science apart from other science and perfected is theology." *Opus Majus*, ch. xix. (trans. Robert Belle Burke, vol. 1, p. 72).

54. For general discussions of the category "natural philosophy," see Ann Blair, "Natural Philosophy," in *The Cambridge History of Science*, vol. 3, *Early Modern Science*, ed. Katherine Park and Lorraine Daston (Cambridge: Cambridge University Press, 2007), pp. 365–406; Andrew Cunningham, "How the *Principia* Got Its Name; or, Taking Natural Philosophy Seriously," *History of Science* 28 (1991): 377–92, and "Getting the Game Right: Some Plain Words on the Identity and Invention of Science," *Studies in History and Philosophy of Science* 19 (1998): 365–89; Peter Dear, "What Is the History of Science the History Of? Early Modern Roots of the Ideology of Modern Science," *Isis* 96 (2005): 390–406; John Heilbron, "Natural Philosophy," in Harrison, Numbers, and Shank, eds., *Wrestling with Nature*, pp. 173–99; C. Lüthy, "What to Do with Seventeenth-Century Natural Philosophy? A Taxonomic Problem," *Perspectives on Science* 8 (2000): 164–95. Simon Schaffer, "Scientific Discoveries and the End of Natural Philosophy," *Social Studies of Science* 16 (1986): 387–420.

55. Ernest Renan, "What Is a Nation?," trans. Martin Thom, in *Nation and Narration*, ed. Homi K. Bhabha (London: Routledge, 1990), p. 11.

56. Karl Deutsch, *Nationalism and Its Alternatives* (New York: Knopf, 1969), p. 19.

Chapter Two

1. Patricia Curd, trans., *Anaxagoras of Clazomenae: Fragments and Testimonia; A Text and Translation* (Toronto: University of Toronto Press, 2007), p. 85.

2. Adolf Harnack, *What Is Christianity?*, trans. Thomas Bailey Saunders, 2nd ed. (New York: G. P. Putnam's Sons, 1908), p. 11.

3. Charles Singer, *A Short History of Science to the Nineteenth Century* [1941] (New York: Dover, 1997), p. 8; J. McClellan and Harold Dorn, *Science and Technology in World History: An Introduction* (Baltimore: Johns Hopkins, 1999), pp. 56, 59–61; Richard Kirby et al., *Engineering in History* (New York: Dover, 1990), p. 42; Lewis Wolpert, "The Unnatural Nature of Science," in *Unveiling the Microcosmos: Essays on Science and Technology from the Royal Institution*, ed. Peter Day (Oxford: Oxford University Press, 1996), pp. 143–56. For ancient authorities on the priority of Thales, see Aristotle, *Metaphysics* 983b20, Isidore of Seville, *Etymologies* II.24.3.

4. Herbert Butterfield, *The Origins of Modern Science: 1300–1800*, 2nd ed. (New York: Macmillan, 1962), pp. vii, 178. See also A. C. Crombie, *Augustine to Galileo*, 2nd ed. (Harmondsworth: Penguin, 1969), p. 24; P. H. Michel, "Greek Science," trans. J. Pomerans, in *Ancient and Medieval Science*, ed. R. Taton (London: Thames & Hudson, 1963), pp. 180–242 (180). H. R. Hall, *The Scientific Revolution, 1500–1800*, 2nd ed. (London: Longmans, 1962), p. 160. For a recent restatement of this view, see Lucio Russo, *The Forgotten Revolution: How Science Was Born in 300 BC and Why It Had to Be Reborn* (Dordrecht: Springer, 2004).

5. This narrative may be found as far back as John Tyndall's "Belfast Address" (1874). John Tyndall, *Fragments of Science*, 2 vols., 8th ed. (London: Longmans, Green,

and Co., 1892), vol. 2, pp. 145–46. Thomas Henry Huxley set out a similar view in *Science and the Christian Tradition* (New York: Appleton, 1894). See also John Burnet's influential *Early Greek Philosophy*, 3rd ed. (London: A. & C. Black, 1920). Burnet contended that Ionian science betrayed "no trace of theological speculation" and continues with the intriguing claim that the ubiquitous references to "god" were "non-religious" (p. 14). Cf. G. E. R. Lloyd, *Early Greek Science: Thales to Aristotle* (London: Norton, 1970); W. K. C. Guthrie, *History of Greek Philosophy*, 6 vols. (Cambridge: Cambridge University Press, 1979), I, 43 f.; S. Sambursky, *The Physical World of the Greeks* (London; Routledge and Kegan Paul, 1956), p. 4; Robert Klee, *Introduction to the Philosophy of Science: Cutting Nature at Its Seams* (Oxford: Oxford University Press, 1996), p. 81; Craig Dilworth, *The Metaphysics of Science* (Dordrecht: Kluwer, 1995), p. 196. A related opposition is that between *mythos* (myth) and *logos* (reason), the latter said to have displaced the former with the rise of Greek philosophy. See Ernst Cassirer, *The Philosophy of Symbolic Forms*, 2 vols., vol. 2, *Mythical Thought*, trans. Ralph Manheim (New Haven: Yale University Press, 1955). But cf. Hans Blumenberg, *Work on Myth*, trans. Robert M. Wallace (Cambridge, MA: MIT Press, 1985).

6. Singer, *A Short History of Science*; Gilbert Murray, *Five Stages of Greek Religion* (London: Watts & Co., 1935); E. R. Dodds, *Pagan and Christian in an Age of Anxiety* (Cambridge: Cambridge University Press, 1965). Peter Gay described this period as moving from "intellectual enquiry to a pathetic hope for salvation." *The Enlightenment: An Interpretation* (New York: Vintage, 1968), p. 118. For a corrective, see Mark Edwards, *Neoplatonic Saints* (Liverpool: Liverpool University Press, 2000), pp. xii–xvi, and Hadot, *What Is Ancient Philosophy?*, pp. 91–97. For a nuanced version of the thesis, see Guy Stroumsa, *The End of Sacrifice: Religious Transformations in Late Antiquity*, trans. Susan Emanuel (Chicago: University of Chicago Press, 2009). And for an application of the thesis to the present, see Stephen Weldon, "In Defence of Science: Secular Intellectuals and the Failure of Nerve Thesis," *Religious Humanism* 30 (1996): 30–39.

7. Popper, "The Myth of the Framework," in *The Myth of the Framework: In Defence of Science and Rationality*, ed. M. A. Notturno (London: Routledge, 1996), pp. 42 f.

8. Paul Davies, *The Goldilocks Enigma: Why Is the Universe Just Right for Life?* (London: Penguin, 2007), p. 14.

9. Bryan Bunch and Alexander Hellemans, *The History of Science and Technology* (Boston: Houghton Mifflin, 2004), pp. 1, 51 f., 81, 93, 142. For similar narratives, see Charles Freeman, *The Closing of the Western Mind* (New York: Vintage, 2005), pp. xviii–xix, 4–6, and passim; Robert Wilson, *Astronomy through the Ages* (Princeton: Princeton University Press, 1997), p. 45; David Deming, *Science and Technology in World History*, vol. 2 (Jefferson, NC: McFarland, 2010), pp. 26–31.

10. Aristotle, *De Anima* 411a9; Euclid, *Elements*, Bk. 1, Prop. 32. See also Sarah Broadie, "Rational Theology," in *The Cambridge Companion to Ancient Greek Philosophy*, ed. A. A. Long (Cambridge: Cambridge University Press, 1999), pp. 205–24; Robin Waterfield, *The First Philosophers* (Oxford: Oxford University Press, 2000), p. xxxiii. On the sacrifice of the ox, see Thomas Heath, tr., *The Thirteen Books of Euclid's Elements*,

3 vols. (New York: Dover, 1956), p. 318. Thales's theorem states that the diameter of a circle subtends a right angle to any point on the circumference.

11. His banishment was probably more because of his political sympathies than his natural philosophical teachings. Curd, *Anaxagoras*, p. 136.

12. For the idea that the Pre-Socratic approach to nature was in many respects a continuation of the preceding mythopoetic worldview, see the classic studies of F. M. Cornford, *From Philosophy to Religion* [1912] (Princeton: Princeton University Press, 1991), pp. 127–30, and Jaeger, *Theology of the Early Greek Philosophers*. See also, more recently, Adam Drozdek, *Greek Philosophers as Theologians: The Divine Arche* (Aldershot: Ashgate, 2007), and David Sedley, who speaks of "a serious misperception of the Presocratic agenda," and points out that the idea "that the world is governed by a divine power is a pervasive assumption of Presocratic thought." *Creationism and Its Critics in Antiquity* (Berkeley: University of California Press, 2007), p. 2. Arguably, Aristotelian teleology is the logical development of this Pre-Socratic tradition. See Edward Engelmann, "Aristotelian Teleology, Presocratic Hylozoism, and 20th Century Interpretations," *American Catholic Philosophical Quarterly* 64 (1990): 297–312.

13. G. E. R. Lloyd, *In the Grip of Disease: Studies in the Greek Imagination* (Oxford: Oxford University Press, 2003), esp. pp. 40–83; and *Demystifying Mentalities* (Cambridge: Cambridge University Press, 1990); Jacques Jouanna, *Hippocrates*, trans. M. B. DeBevoise (Baltimore: Johns Hopkins, 1999); "Hippocratic Oath," *Hippocrates*, LCL 147, p. 298. For a cross-cultural comparison with modern Africa, see Steven Feierman and John M. Janzen, "African Religions," in *Science and Religion around the World*, ed. John Hedley Brooke and Ronald Numbers (Oxford: Oxford University Press, 2011), pp. 229–51.

14. Virtually any recent history of ancient Greek thought, written by a specialist, will give the lie to the idea that science displaces mythology, but see esp. Daryn Lehoux, "Creation Myths and Epistemic Boundaries," *Spontaneous Generations* 3 (2009): 28–34; Keimpe Algra, "The Beginnings of Cosmology," in *The Cambridge Companion to Ancient Greek Philosophy*, ed. A. A. Long (Cambridge: Cambridge University Press, 1999), pp. 45–65, esp. pp. 61–63; Luc Brisson, *How Philosophers Saved Myths* (Chicago: University of Chicago Press, 2004); Richard Buxton, ed., *From Myth to Reason?* (Oxford: Oxford University Press, 1999); Paul Veyne, *Did the Greeks Believe in Their Myths?*, trans. Paula Wissing (Chicago: University of Chicago Press, 1988); Sedley, *Creationism and Its Critics*.

15. Lloyd, *Early Greek Science*, p. xv. Roy Harris, *Semantics of Science* (London, 2005), chs. 1, 2; "Physics," in *Oxford Companion to Classical Civilization*, ed. Simon Hornblower and Anthony Spawforth (Oxford: Oxford University Press, 2004), pp. 532–34.

16. Pierre Hadot, *Philosophy as a Way of Life*, trans. Arnold I. Davidson (Oxford: Blackwell, 1995). See also *What Is Ancient Philosophy?*, trans. Michael Chase (Cambridge, MA: Harvard University Press, 2002). Hadot's *formes de vie* has significant and acknowledged parallels with Wittgenstein's *Lebensformen*. See *Philosophy as a Way of Life*, pp. 17 f., 280; *Wittgenstein et les limites du langage* (Paris: Vrin, 2004).

17. H. Hutter, "Philosophy as Self-Transformation," *Historical Reflections* 16 (1989):

171–98; R. Imbach, "La Philosophie comme exercice spirituel," *Critique* 41 (1985): 275–83; Alexander Nehamas, *The Art of Living: Socratic Reflections from Plato to Foucault* (Berkeley: University of California Press, 1998); John Sellars, *The Art of Living: The Stoics on the Nature and Function of Philosophy* (Aldershot: Ashgate, 2003); Beroald Thomassen, *Metaphysik als Lebensform: Untersuchungen zur Grundlegung der Metaphysik im Metaphysikkommentar Alberts des Grossen* (Münster: Aschendorff, 1985); Michel Foucault, *The Care of the Self*, vol. 3 of *The History of Sexuality*, trans. Robert Hurley (New York: Vintage, 1986), esp. pp. 39–68. Also see John Cottingham, *Philosophy and the Good Life: Reason and the Passions in Greek, Cartesian and Psychoanalytical Ethics* (Cambridge: Cambridge University Press, 1998); Matthew Jones, *The Good Life in the Scientific Revolution: Descartes, Pascal, Leibniz, and the Cultivation of Virtue* (Chicago: University of Chicago Press, 2006).

18. Plato, *Apology* 30a7–b4, 38a. See Terrance Irwin, *Plato's Ethics* (New York: Oxford University Press, 1995), pp. 18 f.; Sellars, *Art of Living*, pp. 33–36; N. Gulley, *The Philosophy of Socrates* (London: Macmillan, 1968), pp. 12–13. Admittedly, Aristotle seems to read Socrates as being preoccupied with definitions. *Eudemian Ethics* 1216b2–25, *Nichomachean Ethics* 1144b18–29. For Cicero, more correctly in my view, the philosophy of Socrates "relates to life and morals." *Tusculan Disputations* 3.4. On this general issue, see Alexander Nehamas, "Socratic Intellectualism," in *Virtues of Authenticity: Essays on Plato and Socrates* (Princeton: Princeton University Press, 1999), pp. 27–58.

19. Plato, *Gorgias* 500 (Hamilton and Cairns ed., p. 283).

20. Sextus Empiricus, *Against the Mathematicians* XI.169.

21. Seneca, *Epistles* 16.3 (LCL 75, p. 105). Cf. *Natural Questions* I, Pref. 13–17.

22. Epictetus, *Discourses* 2.19; 1.15.

23. Diogenes Laertius IX.107.

24. On this theme, see Hadot, *What Is Ancient Philosophy*, pp. 142–44; André-Jean Voelke, *La philosophie comme thérapie de l'âme* (Paris: Cerf, 1993), pp. 107–26. Voelke points out that this is similar to Wittgenstein's image of philosophy (*Tractatus* 6.54) as a ladder that one climbs and then discards.

25. The issue is already present in Socrates's conception of philosophy as an "art" (*technē*), where the art of living is imagined to have both an intellectual component (*logos*) and a practical one (*askēsis*). See Sellars, *Art of Living*, p. 53. Cf. Thomas C. Brickhouse and Nicholas D. Smith, *Plato's Socrates* (Oxford: Oxford University Press, 1994), pp. 5–10.

26. Xenephon, *Memorabilia* 1.1; Aristotle, *Parts of Animals* 642a29; Augustine, *City of God* VIII.3.

27. For an extended treatment of this idea, see Rémi Brague, *The Wisdom of the World: The Human Experience of the Universe in Western Thought*, trans. Teresa Fagan (Chicago: University of Chicago Press, 2003).

28. "The philosopher, holding converse with the divine order, becomes orderly and divine as far as the nature of man allow." Plato, *Republic* VI, 500d, *Collected Dialogues*, 736. And elsewhere: "contact with divine goodness" makes one "god-like." *Laws* X, 904d–e, *Collected Dialogues*, 1460. On wisdom and becoming godlike, see also *Laws*,

IV 716a–d, *Phaedrus* 246d, 248a. *Timaeus* 47c; *Theaetetus* 176a–d; *Republic* X, 613b; Seneca, *Natural Questions* I, Pref. 11–13. Thomas Nagel has recently described Plato's stance in these terms: "Plato was clearly concerned not only with the state of his soul, but also with his relation to the universe at the deepest level. Plato's metaphysics was not intended to produce merely a detached understanding of reality. His motivation in philosophy was in part to achieve a kind of understanding that would connect him (and therefore every human being) to the whole of reality—intelligibly and, if possible, satisfyingly." *Secular Philosophy and the Religious Temperament* (Oxford: Oxford University Press, 2010), p. 4.

29. Plato, *Timaeus* 90d, cf. 47b. See the discussion in Sedley, *Creationism and Its Critics*, pp. 124 f.; Sedley, "The Ideal of Godlikeness," in *Plato*, vol. 2, *Ethics, Politics, Religion, and the Soul*, ed. G. Fine (Oxford: Oxford University Press, 1999), pp. 309–28.

30. Plato, *Laws* 747b. Reference to the motto appears in Olympiodorus, *Prolegomena* 8. Plato was also reputed to have remarked that "God eternally geometrizes." Plutarch, *Quaestiones convivales* 8.2.

31. Aristotle, *Parts of Animals* 656a10–14.

32. "Those who assert that the mathematical sciences say nothing of the beautiful or the good are in error. For these sciences say and prove a great deal about them; if they do not expressly mention them, but prove attributes which are their results or definitions, it is not true that they tell us nothing about them. The chief forms of beauty are order and symmetry and definiteness, which the mathematical sciences demonstrate in a special degree." *Metaphysics* 1078a33.

33. Plato, *Republic* VI, 500d; Aristotle, *Nicomachean Ethics* 1177b20–26.

34. Diogenes Laertius VII.87.

35. "*Secundum naturam vivere*"; Seneca, *Epistles* 95.10 (LCL 77, p. 64).

36. Epictetus, *Discourses* 1.6.

37. Epictetus, *Discourses* 4.7, in *The Works of Epictetus. Consisting of His Discourses, in Four Books, The Enchiridion, and Fragments,* ed. Elizabeth Carter, trans. Thomas Wentworth Higginson (Boston: Little, Brown, and Co., 1865). Socrates relied upon the same logic—that the operations of nature cannot be influenced by us—to argue for the irrelevance of natural philosophy. Xenophon, *Memorabilia* 1.1.

38. Thomas Babington Macaulay, "Lord Bacon," *Edinburgh Review,* July 1837, pp. 1–104 (p. 69).

39. Part of Epicurus's argument was that we should not fear postmortem punishments because, being aggregates of atoms, we will not survive death. Plutarch pointed out, not unreasonably, that many people fear annihilation more than the prospect of future punishments. Plutarch, *That Epicurus actually makes a Pleasant Life Impossible,* [*Moralia*, vol. 14], chs. 25–28.

40. Cicero, *De finibus* I.19.64.

41. Epicurus, *Letter to Herodotus* §37 (= Diogenes Laertius X.37). Cf. Lucretius, *On the Nature of Things* 3, 16 f., 28 f.

42. Diskin Clay, *Paradosis and Survival: Three Chapters in the History of Epicurean Philosophy* (Ann Arbor: University of Michigan Press, 1998), pp. 24, 27.

43. Hadot, *Philosophy as a Way of Life*, p. 83.

44. Ambrose of Milan, *Cain and Abel* 2.22, FC vol. 42, p. 423.

45. Aristotle, *Posterior Analytics* 75a38–39; *Metaphysics* 1025b19–1026a33; *On the Heavens* 299a2–20. See also Amos Funkenstein, *Theology and the Scientific Imagination* (Princeton: Princeton University Press, 1986), pp. 35–37; 303–7.

46. Aristotle, *Metaphysics* 1026a7–10; *Physics* 194a7–11. The term "metaphysics" was coined by a first-century editor. For the most elevated of the sciences Aristotle used the expressions "theology," "first philosophy," "the study of being *qua* being," or simply "wisdom."

47. Aquinas attributes a slightly different pedagogical order to the ancients: "first logic, then mathematics, then natural science, after that moral science, and finally men studied divine science." *Division and Methods of the Sciences* Q. 5, art. 1, 10 (Maurier ed., p. 11). But theology or metaphysics, on account of its elevated subject matter, was always the last science to be attempted: "metaphysics, which deals with divine things, is the last part of philosophy to be learned." SCG I.4.3 (vol. 1, p. 67). Cf. Aristotle, *Nichomachean Ethics* 1142a13–19; Avicenna, *Metaphysics* 1.3 (ed. Van Riet 1:20.77–23.28).

48. Sextus Empiricus, *Against the Logicians* I.16; Diogenes Laertius I.18; VII.39. John Dillon, *The Heirs of Plato: A Study of the Old Academy, 347–274 BC* (Oxford: Oxford University Press, 2003), pp. 98 f.; David Sedley, *Plato's Cratylus* (Cambridge, 2003), pp. 156–59. The Epicureans had a very similar division. Diogenes Laertius X.30.

49. John Sellars, *The Art of Living* (Aldershot: Ashgate, 2003), esp. pp. 78–81. Keimpe Algra, "Stoic Theology," in *Cambridge Companion to the Stoics*, ed. Brad Inwood (Cambridge, 2003), pp. 153–78.

50. Pierre Hadot, *La Citadelle intérieure: Introduction aux Pensées de Marc Aurèle* (Paris: Fayard, 1992), pp. 106–15.

51. Quoted in Hadot, *Philosophy as a Way of Life*, p. 60.

52. Simplicius, *In Physica*, ed. H. Diels, *Commentaria in Aristotelem Graeca* (Berlin: Reimer, 1882), vol. 9, 4.17–5.21, quoted in Brague, *Wisdom of the World*, p. 116. Cf. Brague, *The Legend of the Middle Ages: Philosophical Explorations of Medieval Christianity, Judaism, and Islam*, trans. Lydia G. Cochrane (Chicago: University of Chicago Press, 2009), pp. 77–80.

53. Ptolemy, *Almagest* 1.1. On Ptolemy's religious motivations for pursuing astronomy, see Liba Taub, *Ptolemy's Universe: The Natural Philosophical and Ethical Foundations of Ptolemy's Astronomy* (Chicago and La Salle, IL: Open Court Press, 1993), esp. ch. 5.

54. J. B. Lightfoot, *Commentary on Colossians*, 8th ed. (London, 1875), pp. 154 f.; Johannes Quasten, *Patrology*, 4 vols. (Westminster: Christian Classics, 1986), vol. 1, pp. 251 f. Christian Bunsen has suggested that the recipient may have been the tutor of Roman emperor Marcus Aurelius. *Hippolytus and His Age*, 4 vols. (London: Longman, Brown, Green and Longmans, 1852), vol. 1, p. 188. On date and authorship, see Cyril C. Richardson, ed., *Early Christian Fathers* (New York: Touchstone, 1996), pp. 205–13.

55. *Epistle to Diognetus* 1.1. Greek text from www.earlychurchtexts.com/public /apostfaths/diognetus.html, accessed 10 January 2011. In its earliest manifestation, Christianity in Palestine may be regarded as an eschatological sect within Judaism.

Rudolf Bultmann, *Theology of the New Testament*, 2 vols. (London: SCM, 1965), vol. 1, p. 43. The *Epistle to Diognetus* grapples with the identity of Hellenistic Christianity.

56. For Christianity as a way of life, see also Minucius Felix, *Octavius* 5.1; Eusebius, *Preparation for the Gospel* 1.2. For discussion of the idea of Christian as a "new race," see Denise Kimber Buell, *Why This New Race? Ethnic Reasoning within Early Christianity* (New York: Columbia University Press, 2005), pp. 23 ff.; A. P. Johnson, *Ethnicity and Argument in Eusebius' "Praeparatio Evangelica"* (Oxford: Oxford University Press, 2006); Judith Lieu, *Christian Identity in the Jewish and Graeco-Roman World* (Oxford: Oxford University Press, 2004); Philip Harland, *Dynamics of Identity in the World of the Early Christians* (London: T. & T. Clark, 2009); Matthew Thiessen, *Contesting Conversion: Genealogy, Circumcision, and Identity in Ancient Judaism and Christianity* (Oxford: Oxford University Press, 2011), esp. pp. 143–48. Also see Edwin Judge, "Was Christianity a Religion?," in *The First Christians in the Roman World*, ed. James R. Harrison (Tübingen: Mohr Siebeck, 2008), pp. 404–9.

57. I Tim. 2:10. Respectively, the Wycliffe Bible (1382); the King James Version (1611) and the Douay Rheims New Testament (1582); Young's Literal Translation (1862); Darby translation (1890); New International Version, UK (1983). The Lightfoot translation of the *Epistle to Diognetus* uses "religion." J. B. Lightfoot and J. R. Harmer, trans., and Michael W. Holmes, ed. and rev., *The Apostolic Fathers*, 2nd ed. (Grand Rapids: Baker, 1989), p. 296.

58. Acts 26:5. Young's Literal translation offers "the most exact sect of our *worship*," but the Wycliffe Bible, KJV, Douay-Rheims, Darby, and NIV all use "religion."

59. Col. 2:18.

60. James 1:26, 27. Wycliffe, KJV, Douay-Rheims, Young's, Darby, NIV.

61. There are twenty-one occurrences in the New Testament, ranging over these meanings: race (3); kind (3); kinds (3); family (2); birth (2); countrymen (2); descendant (1); descent (1); nation (1); native (1).

62. There is one occurrence in the Septuagint (Job 14:16) where it means "steps"— as in "following in the footsteps of."

63. 1 Pet. 4:16, Acts 11:26, 26:28. Elias J. Bickerman, "The Name of Christians," *Harvard Theological Review* 42 (1949): 109–24.

64. Gal. 3:28, NIV.

65. Maurice Sachot, "Comment le Christianisme est-il devenu *religio*," *Revue des sciences religieuses* 59 (1985): 95–118; "«Religio/Superstitio», Historique d'une subversion et d'un retournement," *Revue d'histoire des religions* 208 (1991): 355–94. Denise Kimber Buell, "Race and Universalism in Early Christianity," *Journal of Early Christian Studies* 10 (2002): 429–68.

66. Jn. 4:23. Ambrose of Milan speaks of this new kind of worship as "true religion" (*vera religio*), which he contrasts with superstition. *De excidio urbis Hierosolymitanae libri quinque*, PL 15, col. 2084B.

67. Seth Schwartz, *Imperialism and Jewish Society from 200 B.C.E. to 640 C.E.* (Princeton: Princeton University Press, 2001), p. 179; Mary Beard, John North, and Simon Price, *The Religions of Rome*, 2 vols. (Cambridge: Cambridge University Press, 1998),

vol. 1, p. 43; Charles Taylor, *A Secular Age*, pp. 146–58. Cf. Brent Nongbri, "Dislodging 'Embedded' Religion: A Brief Note on a Scholarly Trope," *Numen* 55 (2008): 440–60.

68. I Cor. 1:22; Gal. 3:28.

69. For a different version of supercessionism in relation to Judaism, see K. Randall Soulen, *The God of Israel and Christian Theology* (Minneapolis: Fortress Press, 1996).

70. On these relations, see Adam H. Beck and Annette Yoshiko Reed, eds., *The Ways That Never Parted: Jews and Christians in Late Antiquity and the Early Middle Ages* (Minneapolis: Fortress Press, 2007); Daniel Boyarin, *Border Lines: The Partition of Judeao-Christianity* (Philadelphia: University of Pennsylvania Press, 2004); Guy Stroumsa, *The End of Sacrifice: Religious Transformations in Late Antiquity*, trans. Susan Emanuel (Chicago: University of Chicago Press, 2009).

71. Stroumsa, *The End of Sacrifice*, p. 65. Stroumsa also speaks of a "new model of religion, in which authority is no longer exterior and public, but rather interior and internalized, whether in the self or the sacred Book" (p. 92). He identifies four such "transformations" adding to the two mentioned here: the communitarianism of the Jewish Diaspora and the invention of a new religious "self" that was to be the subject of scrutiny and training.

72. Ibid., pp. 39, 88.

73. Books with significant intellectual content were typically read aloud in antiquity, although silent reading was not unknown. See Brian Stock, *Augustine the Reader: Meditation, Self-Knowledge and the Ethics of Interpretation* (Cambridge, MA; Harvard University Press, 1996), p. 5. Arguably, with word separation and punctuation, the codex facilitated silent reading. See Robert A. Kaster, *Guardians of Language: The Grammarian and Society in Late Antiquity* (Berkeley: University of California Press, 1997), pp. 32–50.

74. Daniel Boyarin has suggested that "Judaism" was a Christian invention. "The Christian Invention of Judaism," in *Religion: Beyond a Concept*, ed. Hent de Vries (New York: Fordham University Press, 2008), pp. 150–77. See also Schwartz, *Imperialism and Jewish Society*, pp. 179–202. On the terms "Jews" (*Ioudaios*) and "Judaism" (*Ioudaismos*), see Shayne J. D. Cohen, *The Beginnings of Jewishness: Boundaries, Varieties, Uncertainties* (Berkeley: University of California Press, 1999), pp. 69–106.

75. Tertullian, *De Praescriptione Haereticorum* 7. Celsus quoted in Origen, *Contra Celsum* III, 55; I, 9, ed. and trans. Henry Chadwick (Cambridge: Cambridge University Press, 1965), pp. 165, 12. See also Peter Harrison and David C. Lindberg, "Early Christianity," in Brooke and Numbers, *Science and Religion around the World*, pp. 67–91.

76. Eusebius, *Preparation for the Gospel* 1.3, trans. E. H. Gifford (Oxford: Clarendon, 1903), p. 7.

77. Eusebius, *Church History* 19.6 (NPNF 2, vol. 1, p. 265).

78. Eusebius, *Preparation* 1.5.10 (trans. Gifford, p. 19).

79. Augustine, Sermon 150, 4 in *Works*, vol. III/5, p. 31. The "five different sets of special opinions" were the philosophical schools — Academics, Peripatetics, Epicureans, Stoics, and Cynics. See also *The Trinity* XIII.7.10; *City of God* VIII.3; XVIII.41; XIX.1, *Of True Religion* ii.2. Cf. Athenagoras, *A Plea for the Christians* 7.

80. Aquinas, SCG 3.48.14–15 (vol. 3/1, p. 167). See also Kerr, *After Aquinas*, pp. 65 f. In SCG I.4.3, Aquinas also points out that "almost all of philosophy is directed towards the knowledge of God" (vol. 1, p. 67).

81. Justin, *Dialogue with Trypho*, 2 (ANF vol. 1, p. 195). Justin was less impressed with the "physical" philosophy of Thales and his school (*Hortatory Address to the Greeks*, 3).

82. *Stromata* I.1 (ANF vol. 2, p. 303), *Stromata* I.v (ANF vol. 2, p. 305). "The studies of philosophy, therefore, and philosophy itself, are aids in treating of the truth. . . . Accordingly the soul must be prepared and variously exercised, if it would become in the highest degree good. For there is the scientific and the practical element in truth; and the latter flows from the speculative; and there is need of great practice, and exercise, and experience." *Stromata* XI (ANF vol. 2, p. 501).

83. Origen, *Letter from Origen to Gregory* 1.

84. Augustine, *Contra Academicos* III.ixx.42; *Contra Julianum* IV.72; *Contra Julianum opus imperfectum* IV.22; *De vera religione* I.i.5. Letter 137, Augustine to Volusian 5, 17. For Augustine on pagan science, see David C. Lindberg, "The Medieval Church Encounters the Classical Tradition: Saint Augustine, Roger Bacon, and the Handmaiden Metaphor," in Lindberg and Numbers, *When Science and Christianity Meet*, pp. 7–32. On Christianity as true philosophy, see Sachot, *Quand le christianisme a changé le monde*, esp. p. 260.

85. Diels and Kratz, *Die Fragmente der Vorsokratiker*, 3 vols., 11th ed. (Zurich: Wiedmann, 1964), 59B 21, vol. 2, p. 43, trans. in Brague, *The Legend of the Middle Ages*, p. 80. Cf. Aristotle: "to gain light on things imperceptible we must use the evidence of sensible things," *Nichomachean Ethics* 1104a13–14. For a brief discussion, see James Allen, *Inference from Signs: Ancient Debates about the Nature of Evidence* (Oxford: Oxford University Press, 2001), pp. 1–12. The Pauline reference comes from Rom. 1:20 (NIV).

86. Origen, *In Cantica Canticorum: Prologus* 3 PG, 13, 73–75 (ACW, vol. 26, pp. 39–42); cf. Clement, *Stromata* 1.28; IX. Robert M. Berchman, *From Philo to Origen: Middle Platonism in Transition* (Chico: Scholars Press, 1984), p. 140; Paul Olsen, *The Journey to Wisdom: Self-Education in the Patristic and Medieval Literature* (Lincoln, 1995), pp. 16–41.

87. See Jean Daniélou, *Platonisme et théologie mystique*, 2nd ed. (Paris: Editions Montaigne, 1953), pp. 17–26; Mariette Canevet, *Gregoire de Nysse et l'herméneutique biblique* (Paris; Études Augustiniennes, 1983), p. 127. Evagrius has a similar division of knowledge, *Schol. In Prov.* 247 (*Sources chrétiennes* 340–42). See Jeremy Driscoll's introduction to *Evagrius: Ad monarchos* (ACW vol. 59, p. 18); Paul M. Blowers, "'Entering This Sublime and Blessed Amphitheatre': Contemplation of Nature and Interpretation of the Bible in the Patristic Period," in *Nature and Scripture in the Abrahamic Traditions, vol. 1, To 1700*, ed. S. Mandelbrote and J. van de Meer (Leiden: Brill, 2008), pp. 147–77 (esp. p. 162).

88. John Cassian, *The Conferences* III.vi.4 (ACW vol. 57, p. 125).

89. In God "are to be found the cause of existence, the ultimate reason for the understanding, and the end in reference to which the whole life is to be regulated. Of which three things, the first is understood to pertain to the natural, the second to the rational, and the third to the moral part of philosophy." Augustine, *City of God* VIII.4

(Dodds trans., p. 248). See also Pierre Hadot, "Les divisions des parties de la philosophie," *Museum Helveticum* 36 (1978): 201–23.

90. Augustine, Letter 137, Augustine to Volusian 5, 17 (*Works*, vol. II/2, pp. 222–23).

91. Augustine, *The Trinity* XIII.6 (*Works*, vol. I/5, pp. 363–64.). In the thirteenth
century, Robert Grosseteste (ca. 1175–1253) takes up Augustine's view, discussing
Augustine's dictum that "whatever one has learned outside of this scripture, if it is
harmful, then it is condemned in the text, if it is useful then it is found there." Grosseteste argues that natural philosophy assists in the interpretation of scripture, and
that the proper exposition of scripture leads to instruction in faith or the building up
of love. This is the goal of sacred science, and natural philosophy assists insofar as it
helps achieve these goals. The reading of scripture leads to "the edification of habitual
thinking" (*edificatio morum*). See James R. Ginther, *Master of the Sacred Page: A Study of
the Theology of Robert Grosseteste, ca. 1229/30–1235* (Aldershot: Ashgate, 2004), pp. 68,
69.

92. Ignatius, *Epistle to the Magnesians* 8.2.

93. Tatian, *Address to the Greeks* 31. Cf. Justin Martyr, *Apology* 1.46; Minucius Felix,
Octavius 20, 1; Clement of Alexandria, *Stromata* 1.4.28.

94. Eusebius, *Proof of the Gospel* 1.2. Rendering *theosebeia* as "godliness," and
eusebeia as "piety." Cf. *Ecclesiastical History* 1.4.1.

95. Augustine, Letter 102, Augustine to Deogratias 21, 15, *Works*, vol. II/2, pp. 31, 28.

96. Samuel Pufendorf would thus remark in 1687 that "the Christian religion is,
by God's peculiar providence, endowed with such Qualifications, that it consequently
deserves the Name of an Universal Religion." This, he goes on to say, is because it is
not associated with a particular place or temple, has no hereditary priesthood, is not
linked to a particular nation, and transcends ethnicity and gender. He makes explicit
reference to the "neither Jew nor Greek" passages. *Of the Nature and Qualification of
Religion in Reference to Civil Society*, ed. Simone Zurbuchen (Indianapolis: Liberty Fund,
2001), pp. 26 f.

97. 2 Tim. 1:13; 2 Tim. 4:3; Titus, 1:9; 1 Tim. 6:20; John 1:1. For further examples, see
J. N. D. Kelly, *Early Christian Creeds*, 3rd ed. (London: Continuum, 2006), pp. 7–11.

98. Seneca states the latter problem thus: "Men say: 'The happy life consists in
upright conduct; precepts guide one to upright conduct; therefore precepts are sufficient for attaining the happy life.'" *Epistles* 95.4 (LCL 77, p. 61). For Seneca "doctrines" (*decreta*) include the Greek *dogmata*, and also the Latin *scita* and *placita*. For
a discussion of Seneca on this question, see Paul R. Kolbet, *Augustine and the Cure of
Souls: Revising a Classical Ideal* (Notre Dame: University of Notre Dame Press, 2010),
pp. 41–64.

99. Seneca, *Epistles* 95.12 (LCL 77, pp. 65–67).

100. SV 41, trans. in Clay, *Paradosis and Survival*, p. 27.

101. Augustine, *Sermon* 58.11, trans. in Kelly, *Early Christian Creeds*, p. 370.

102. Augustine, *Sermon* 58.13 (*Works*, vol. III/3, pp. 124 f.) A number of the Church
Fathers traced the source of the Delphic maxim to Moses or Solomon (Exod. 10:28,

34:12; Deut. 4:9; Song of Songs 1:8), contending that the Greeks had plagiarized it. See, e.g., Clement, *Stromata* 2.15.71; Ambrose of Milan, *Expositions of the Psalms* 118, 2.13.

103. Augustine, *A Treatise on the Faith and the Creed* 10.25 (NPNF I, vol. 3, p. 333). He speaks here of the Apostle's Creed. This idea will later be incorporated into Anselm's famous motto *fides quaerens intellectum* (faith seeking understanding).

104. "*Nisi credideritis non intelligetis.*" The saying is taken from the Septuagint version of Isaiah 7:9. The Hebrew reads rather differently.

105. Augustine, *Sermon* 43.1 (*Works*, vol. III/3, p. 238). Cf. *On Christian Doctrine* 2.12.17; *The Trinity* VII.4.12.

106. Plato, *Charmides* 157a (Hamilton and Cairns, p. 103). The curative power of words is perhaps a vestige of the "magical" qualities possessed by language in archaic Greece. See Marcel Detienne, *The Masters of Truth in Archaic Greece*, trans. Janet Lloyd (New York: Zone, 1999). See also Gaukroger, *Emergence of a Scientific Culture*, pp. 229 f. The "religious" conception of truth, Detienne argues, gives way to the more secular understanding represented in philosophy. But rhetoric, which is in some tension with philosophy, retained some of the causal powers of words.

107. Frances Young, "The Rhetorical Schools and Their Influence on Patristic Exegesis," in *The Making of Orthodoxy: Essays in Honour of Henry Chadwick*, ed. Rowan Williams (Cambridge: Cambridge University Press, 1989), pp. 182–99; A. D. Nock, *Conversion* (Oxford: Oxford University Press, 1933), p. 177; Kolbet, *Augustine and the Cure of Souls*, pp. 1–12.

108. Cicero, *Tusculan Disputations* 5.27.28 (LCL 141, 504–6).

109. Augustine wrote that there are two penalties for every sinful soul: "ignorance and difficulty." *De Libero Arbitrio* 3.20.55. Aquinas spoke of four consequences of sin: ignorance, malice, weakness, concupiscence. ST 1a2ae, 61, 2; 1a2ae, 85, 3.

110. Clement of Alexandria, *Stromata* 7.4 (ANF vol. 2, p. 529). On the connection between immorality and heresy, see R. M. Grant, "Charges of 'Immorality' against Various Religious Groups in Antiquity," in *Studies in Gnosticism and Hellenistic Religions*, ed. R. van den Broek and J. Vermaseren (Leiden: Brill, 1981), pp. 161–70.

111. "Receipt of Constantine to the Bishops of Numidia," in *The Work of St. Optatus*, trans. O. R. Vassall-Phillips (London: Longmans, Green and Co., 1917), p. 413.

112. For treatments of the relevant terms, see G. Kittel, ed., *Theological Dictionary of the New Testament* (Grand Rapids: Eerdmans, 1968), s.v. "πιστεύω," etc., vol. 6, pp. 174–228; Smith, *Faith and Belief*; Berard L. Marthaler, *The Creed: The Apostolic Faith in Contemporary Theology*, 3rd ed. (New London: Twenty-Third Publications, 2007), pp. 18–36.

113. Charles T. Lewis and Charles Short, *A Latin Dictionary* (Oxford: Oxford University Press, 1879), s.v. "credo."

114. Seneca, *Epistles* 95.50 (LCL 77, p. 89). "Whoever imitates them, is worshipping them sufficiently." Ibid. (LCL 77, p. 91).

115. "*Quid est ergo credere in eum? Credendo amare, credendo diligere, credendo in eum ire, et eius membris incorporari.*" PL 35, col. 1631. English translation in Augustine, *Homilies on the Gospel of John*, Tractate 29, 6 (NPNF I, vol. 7, p. 185); *Expositions of the Psalms*

77, 8 (*Works*, vol. III/18, p. 98). See also Augustine, Sermon 14a, 3, in *Works*, vol. III/11, p. 26.

116. "Believe" verb. 1. "To have confidence or faith *in* (a person), and consequently to rely upon, trust to. Const. *in*, and (in theological language) *on* (*an* obs.); formerly with *into, unto, of* (rare)." *OED*, s.v. "believe," http://www.oed.com/view/Entry/17376 ?redirectedFrom=believe#eid, accessed 10 May 2013.

117. "*In Deum ergo credere, hoc est fideliter eum quaerere, est iota in eum dilectione transire. Credo ergo in illum, hoc est dicere, confiteor illum, colo illum, adore illum, totum me is jus ejus ac dominum trado alque transfundo.*" Faustus of Riez, *De spiritu sancto* I.1 (PL 62, 10c–d), trans. in Henri de Lubac, *The Splendor of the Church* (San Francisco: Ignatius Press, 1999), p. 35. The confession of faith, thus understood, is in some respects akin to what contemporary philosophers refer to as a "performative utterance." See J. L. Austen, *Doing Things with Words* (Oxford: Oxford University Press, 1962).

118. The *OED* dates the first uses of "believe in" in the sense of "To have confidence in or be convinced of the actual existence or occurrence of a thing," from 1659. *OED*, s.v. "Believe," 7 b. http://www.oed.com/view/Entry/17376?redirectedFrom=believe&, accessed 10 May 2013. For an insightful discussion of these changes in meaning and their significance, see Smith, *Faith and Belief*, pp. 105–27. See also R. Needham, *Belief, Language, and Experience* (Oxford: Blackwell, 1972); J. Pouillon, "Remarks on the Verb 'to Believe,'" in *Between Belief and Transgression: Structural Essays in Religion, History, and Myth*, ed. M. Izard and P. Smith (Chicago: University of Chicago Press, 1982); M. Ruel, "Christians as Believers," in *Religious Organization and Religious Experience*, ed. J. Davis (London: Academic Press, 1982).

119. Rodney Needham, *Circumstantial Deliveries* (Berkeley: University of California Press, 1982), p. 78. The argument for this thesis is set out in detail in *Belief, Language and Experience*.

120. Lewis and Short, *A Latin Dictionary*, s.v. "doctrina."

121. Augustine, *Retractions* 2.4. See also I. Opelt, "Doctrina und doctrina christiana," *Der altsprachliche Unterricht* 9 (1966): 5–22; G. A. Press, "Doctrina in Augustine's *De doctrina christiana*," *Philosophy and Rhetoric* 17 (1984): 98–120.

122. Bonaventure, *Brevil. Prol.* Q. 5, 201; Aquinas, *In Beot de trin* Q. 5, Q. 4. Hugh of Saint Victor identifies mathematics with *doctrina*. *Didascalicon* 2.1, 2.3 (PL 175, 753). Mathematics was the "doctrinal science" (*doctrinalis scientia*) because it represents the first degree of abstraction from corporeal things and thus prepared the way for the study of the wholly immaterial objects of theology.

123. Hadot, *La Citadelle intérieure*, pp. 51–68; Stroumsa, *End of Sacrifice*, p. 52.

124. Augustine, *City of God* XIX.1. Cf. XVIII.41. Augustine's source, Varro's *De philosophia*, is no longer extant. On the diversity of philosophical teachings, see also Basil, *Hexameron* 1.11; Tertullian, *Treatise on the Soul* 3; *Ad nations* 2.1. In response to this standard criticism, "concordists" such as Simplicius sought to demonstrate a fundamental agreement among the philosophers, and Plato and Aristotle in particular. See Constance Blackwell, "Neo-Platonic Modes of Concordism versus Definitions of Differ-

ence," in *Laus Platonici Philosophi: Marsilio Ficino and His Influence,* ed. Stephen Clucas, Peter J. Forshaw, and Valery Rees (Leiden: Brill, 2011), pp. 317–42.

125. Cicero, *De divinatione* 2.58.119.

126. Augustine, *On Christian Doctrine* 4.28.61.

127. I Cor. 8:1. Augustine, *On Christian Doctrine* 1.36.40; 3.15.23 (*regnum caritatis* — trans. "reign of love" in NPNF I); 31.39.43. Elsewhere in this work, Augustine explains that scripture "is a narrative of the past, a prophecy of the future, and a description of the present. But all these tend to nourish and strengthen charity, and to overcome and root out lust." 3.10.15.

128. "*In nobismetipsis namque debemus tranformare quod legimus; ut cum per audiaum se animus excitat, ad operandum quod audierit vita concurrat.*" Gregory, *Moralia in Job* 1.33, PL 75, 542.

129. Brian Stock, *Augustine the Reader*, p. 12. See also Stock's *After Augustine: The Meditative Reader and the Text* (Philadelphia: University of Pennsylvania Press, 2001), passim.

130. One of the few weaknesses of W. C. Smith's account of religion in the West is that he fails to provide an adequate account of the place of doctrinal statements in patristic and medieval Christianity.

131. Much the same is true for other cultures. See Brooke and Numbers, *Science and Religion around the World*, esp. pp. 1–5, 229–32, 278–80.

132. For some additional and general critiques of the Popper-style narrative, see the essays of David C. Lindberg and Michael H. Shank in *Galileo Goes to Jail, and Other Myths about Science and Religion,* ed. Ronald L. Numbers (Cambridge, MA: Harvard University Press, 2010), pp. 8–18, 19–27; Harrison and Lindberg, "Early Christianity."

133. Aristotle, *Metaphysics* 1026a21; Chrysippus in *Stoicorum Veterum Fragmenta,* ed. Hans von Arnim (Stutgardiae: B. G. Teubneri, 1964), vol. 2, p. 42.

134. This is already explicit in Xenophanes's critique of the anthropomorphic polytheism, and his proposed substitute for it, the "supreme God . . . who is not like mortals in body or in mind." Zeller, *Vorsokrastische Philosophie*, pp. 524, 525, 530, n. Consider also that Plutarch attributes to Anaxagoras the fact that Pericles abandoned superstition in relation to the heavenly bodies, and embraced true piety. H. Diels and W. Kranz, eds., *Die Fragmente der Vorsokratiker,* 3 vols., reprint of 6th ed. (Berlin: Weidmann, 1974), 59 A 16.

135. Origen, *Contra Celsum* 5.7. Cf. Proclus, *Commentary on the Cratylus* 125. On ancient belief in the divinity of the heavens, see Alan Scott, *Origen and the Life of the Stars* (Oxford: Clarendon, 1991), p. 55 and passim. Justin Martyr, *First Apology* ch. 6: "Thus we are even called atheists. We do proclaim ourselves atheists as regards those whom you call gods, but not with respect to the most true God" (FC, vol. 6, pp. 38–39). J. J. Walsh, "On Christian Atheism," *Vigiliae Christianae* 45 (1991): 255–77; R. Jungkuntz, "Fathers, Heretics and Epicureans," *Journal of Ecclesiastical History* 17 (1966): 3–10; A. D. Simpson, "Epicureans, Christians, Atheists in the Second Century," *Transactions and Proceedings of the American Philological Association* 72 (1941): 372–81. For parallels

between Christianity and Epicureanism, see Robert R. Gregg, *Consolation Philosophy: Greek and Christian Paideia in Basil and the Two Gregories* (Cambridge, MA: Philadelphia Patristic Foundation, 1975); Paul A. Holloway, "Bona Cogitare: An Epicurean Consolation in Phil 4:8–9," *Harvard Theological Review* 91 (1998): 89–96.

136. Philoponus *contra Aristotelem ap.* Simplicium *in Cael.* 88, 28–34, in Richard Sorabji, *The Philosophy of the Commentators 200–600 AD*, vol. 2, *Physics* (London: Duckworth, 2003), p. 374.

137. C. Wildberg, "Impetus Theory and the Hermeneutics of Science in Simplicius and Philoponus," *Hyperboreus* 5 (1999): 107–24; C. Scholten, *Antike Naturphilosophie und christliche Kosmologie in der Schrift "De opificio mundi" des Johannes Philoponos* (Berlin: De Gruyter, 1996); Richard Sorabji, ed., *Philoponus and the Rejection of Aristotelian Science* (Ithaca: Cornell University Press, 1987); William A. Wallace, *Galileo and His Sources: The Heritage of the Collegio Romano in Galileo's Science* (Princeton: Princeton University Press, 1984).

138. See, e.g., Tertullian, *Ad nationes* 2.6; Augustine, *City of God* VIII.18–26; *On Christian Doctrine* 2.20–23.

Chapter Three

1. Ambrose of Milan, *Hexameron* 1.4 (FC vol. 42, pp. 15 f.).

2. Augustine, *On Christian Doctrine* 1.4 (NPNF I, vol. 2, p. 523).

3. Roger Bacon, *Opus Majus*, 2 vols., trans. Robert Belle Burke (Philadelphia: University of Pennsylvania Press, 1928), vol. 2, p. 636.

4. Thomas Nagel, "Secular Philosophy and the Religious Temperament," in his *Secular Philosophy and the Religious Temperament* (Oxford: Oxford University Press, 2010), pp. 3–17, quoted from pp. 3, 4.

5. From the extensive literature on the "book of nature" topos, see G. Tanzella-Nitti, "The Two Books Prior to the Scientific Revolution," *Annales Theologici* 18 (2004): 51–83; Klaas van Berkel and Arie Vanderjagt, eds., *The Book of Nature in Antiquity and the Middle Ages* (Louvain: Peeters, 2005) and *The Book of Nature in Early Modern and Modern History* (Louvain: Peeters, 2006); Peter Harrison, *The Bible, Protestantism and the Rise of Natural Science* (Cambridge: Cambridge University Press, 2001); Jitse van der Meer and Scott Mandelbrote, eds., *Nature and Scripture in the Abrahamic Religions*, 2 vols. (Leiden: Brill, 2008).

6. Origen, *The Song of Songs, Commentary and Homilies*, trans. R. P. Lawson (London: Longmans, Green and Co., 1957) ACW vol. 26, pp. 218 f., 223; Origen, *On First Principles* 4.1.11. Paul M. Blowers, "'Entering This Sublime and Blessed Amphitheatre': Contemplation and Interpretation of Nature in the Patristic Period," in van der Meer and Mandelbrote, *Nature and Scripture in the Abrahamic Religions: Up to 1700*, vol. 1, pp. 147–76 (152); Francis Young, *Biblical Exegesis and the Formation of Christian Culture* (Cambridge: Cambridge University Press, 1997), pp. 186–213.

7. Athanasius, *Contra gentes* §34 (NPNF II, vol. 4, p. 22); Chrysostom, *Homily* 9.5 (NPNF 1, vol. 9, pp. 401–2). Augustine, *Contra Faustum Manichaeum* 32.20.

8. Evagrius, *Scholia in Psalmos* 138.16. Cf. Maximus the Confessor, *Questions to Thalassius* 32.

9. Origen's threefold scheme is set out in *On First Principles* 4.1.11 (ANF vol. 4, p. 359); *Homilies on Leviticus* 5.3 (FC vol. 83, p. 89). Cf. Clement of Alexandria, *Stromata* 1.28; Jerome, *Commentariorum in Ezechielem* 5.16.30–31 (PL 25, 153D); *In Amos* II, 4.4–6 (PL 25, 1027D –1028A). Hugh of Saint Victor also spoke of three senses: historical, allegorical, and tropological. *Didascalicon* 5.2; *De Sacracmentis* Prologue 4 (PL 176, 184–85). Augustine distinguishes these four senses: history, etiology, analogy, and allegory. *De utilitate credendi* 5; *De Genesi ad litteram imperfectus liber* II (PL 34, 222). The more common fourfold division may be found in Gregory the Great, *Homilia in Ezechielem* 2.9.8 (PL 76, 1047B); Eucherius, *Liber Formularum spiritualis intelligentiae*, preface (PL 50, 727 f.); Rabanus Maurus, *Ennarationes in epistolas Pauli* 15.4 (PL 122, 331); Bonaventure, *Itinerarium* 4.6. For a comprehensive overview, see Henri de Lubac, *Medieval Exegesis: The Four Senses of Scripture*, trans. E. M. Macierowski, 3 vols. (Grand Rapids: Eerdmans, 1998–2009).

10. Cassian, *The Conferences* 14.8.4 (ACW vol. 57, p. 510). The passage is Gal. 4:24, which itself makes reference to the allegorical interpretation of Gen. 16:15, 21:2, 9.

11. Hugh of Saint Victor, *De Sacracmentis*, Prologue 4 (PL 176, 184–85); Aquinas, ST 1a, 1, 10. An alternative division, observed by Bede, links history and tropology with "things below," and allegory and anagogy with "things above." De Lubac, *Medieval Exegesis*, vol. 2, pp. 1–39.

12. Hugh of Saint Victor, *De tribus diebus* 4 (PL, 122, 176.814 B–C); *Didascalicon* 5.3, trans. Jerome Taylor (New York: Columbia University Press, 1991), p. 121; cf. *De sacramentis* I. Prologus 5 (PL 76, 185C), *Didascalicon* III. 19 (p. 101). This last passage was a favorite of cultural critic Edward Said.

13. Bonaventure, *Itinerarium mentis in deum* 1.12, 13 in *The Works of Bonaventure*, vol. 1, pp. 26 f.

14. Aquinas, ST Ia, 1, 10. Cf. *Commentary on St. Paul's Epistle to the Galatians* Ch. 4 Lec. 7, p. 137. Some scholars have suggested that Thomas's exegesis represents a break with the Origenist-Augustinian approach. See, e.g., Beryl Smalley, *The Study of the Bible in the Middle Ages*, 2nd rev. ed. (New York: Philosophical Library, 1952), pp. xv, 41, 263, 292–94, 300–302. But cf. De Lubac, *Medieval Exegesis*, vol. 3, esp. pp. 1–5. Roger Bacon manages to link the two traditions by suggesting that natural philosophy will "present the literal truth of scripture most effectively, so that through suitable adaptations and similitudes the spiritual sense may be derived." *Opus Majus*, 2 vols., trans. Robert Belle Burke (Philadelphia: University of Pennsylvania Press, 1928), vol. 2, p. 631. The natural philosophy that enables us to do this Bacon labels "experimental science."

15. Although for Augustine the verbal and the causal are linked, because the divine word has causal effects, and indeed is eternally causally efficacious. *De civitate Dei* xvi.vi.1.

16. *Physiologus* vi, from *Physiologus: A Medieval Book of Nature Lore*, trans. Michael Curley (Chicago: University of Chicago Press, 1979), pp. 9 f.

17. Aberdeen University Library MS 24, fol. 35r. Trans. at http://www.abdn.ac.uk
/bestiary/translat/35r.hti, accessed 28 January 2011.

18. The third verse of Thomas's hymn reads: "*Pie Pellicane, Jesu Domine, / Me immun-
dum munda Tuo sanguine*" (Loving pelican, Oh Jesus, Lord / Unclean am I but cleanse
me in your blood). For the pelican in art, architecture, and literature, see Emile Mâle,
Religious Art in France: The Twelfth Century, trans. Marthiel Mathews (Princeton: Princeton University Press, 1978), ch. 9; Arthur Collins, *Symbolism of Animals and Birds Represented in English Church Architecture* (London: Pitman and Sons, 1913), pp. 33 f.; Louis
Charbonneau-Lassay, *The Bestiary of Christ* (New York: Arkana, 1972), pp. 8 f. For references to the pelican in Dante and Shakespeare, see *Paradiso*, xxv.113, *King Lear*, III.iii.75.
Thomas Browne treats the story of the pelican as a vulgar error. *Pseudodoxia Epidemica*
IV.v, ed. Robin Robbins, 2 vols. (Oxford: Clarendon, 1981), vol. 1, pp. 366–69; also see
the comprehensive commentary provided by Browne's editor (II, 946–48).

19. On this theme, see Harrison, *Bible and the Rise of Science*, pp. 21–23.

20. Ambrose of Milan, *Hexameron*, trans. John Savage, FC vol. 42 (New York: Fathers
of the Church, 1961), 5.21.70 (p. 215); 5.15.50 (p. 200); 5.5. (p. 168); 3.13 (p. 106).

21. Basil the Great, *Hexameron* 1.6, in *Saint Basil: Exegetic Homilies*, trans. Agnes
Way, *Fathers of the Church* 46 (Washington, DC: Catholic University of America Press,
1963), p. 11.

22. One alternative explanation was the principle of plenitude, according to which
the best cosmos will exhibit the greatest variety of creatures. A. O. Lovejoy, *The Great
Chain of Being* (Cambridge, MA: Harvard University Press, [1936] 2001).

23. Richard of Saint Victor, *The Mystical Ark* 2.14, in *The Twelve Patriarchs, the Mystical Ark, Book Three of the Trinity*, trans. Grover A. Zinn (New York: Paulist Press, 1979),
p. 194. Cf. Augustine: "Every sign is also a thing . . . but not every single thing is also a
sign." *De doctrina christiana* 1.2.2.

24. Ambrose, *Hexameron* 3.9 (p. 96).

25. Basil, *Hexameron* 1.6, in *Saint Basil: Exegetic Homilies*, trans. Agnes Way, *Fathers
of the Church* 46 (Washington, DC: Catholic University of America Press, 1963), p. 11.
Both Basil and Ambrose link the uses of things to Paul's explanation in Rom. 1:20 of
the relation between the visible and invisible things. Ambrose, *Hexameron* 1.4; Basil,
Hexameron 1.6.

26. Aristotle, *Politics* 1.8. 1256b. David Sedley, "Is Aristotle's Teleology Anthropocentric?," *Phronesis* 36 (1991): 179–96; John Passmore, *Man's Responsibility for Nature*
(London: Duckworth, 1974), Part I. See also Clarence Glacken, *Traces on the Rhodian
Shore: Nature and Culture in Western Thought from Ancient Times to the End of the Eighteenth Century* (Berkeley: University of California Press, 1973); Peter Harrison, "Subduing the Earth: Genesis 1, Early Modern Science, and the Exploitation of Nature," *Journal
of Religion* 79 (1999): 86–109. But cf. Lynn White Jr., "The Historical Roots of our Ecological Crisis," *Science* 155 (1967): 1203–7.

27. On this distinction, see W. R. O'Connor, "The *Uti-frui* Distinction in Augustine's
Ethics," *Augustinian Studies* 14 (1983): 45–62; Harrison, "Reinterpreting Nature in Early
Modern Europe: Natural Philosophy, Biblical Exegesis, and the Contemplative Life," in

The Word and The World: Biblical Exegesis and the Emergence of Modern Science, ed. K. Killeen and P. Forshaw (London: Palgrave Macmillan, 2007), pp. 25–44.

28. Augustine, *De doctrina christiana* 1.4.4 (NPNF 1, vol. 2, p. 523). Cf. *City of God* XV.7.

29. Augustine, *De doctrina christiana* I.27.28; *City of God* XV.22. For Augustine, preoccupation with the creatures is linked to the intellectual vice of curiosity. See Peter Harrison, "Curiosity, Forbidden Knowledge."

30. Aquinas, ST 2a2ae, 9, 4. Cf. Peter Lombard, *Sentences* I.1.2.

31. Aquinas, ST 2a2ae, 180, 4. Here Aquinas cites Augustine as an authority: "Hence Augustine says (*De Vera Relig.* xxix) that 'in the study of creatures we must not exercise an empty and futile curiosity, but should make them the stepping-stone to things unperishable and everlasting.'" He also makes allusion to Richard of Saint Victor's six stages of contemplation, the first three of which involve the contemplation of "corporeal things," concluding that "the contemplation of truth regards not only the divine truth, but also that which is considered in creatures."

32. Bonaventure, *Itinerarium mentis in deum* 1.7, *Works of Bonaventure*, vol. 1, p. 11.

33. Bonaventure, *Collationes in Hexaëmeron* 12, quoted in Brague, *Legend of the Middle Ages*, pp. 80 f.

34. Hugh of Saint Victor, *Didascalicon* 2.i (p. 61).

35. Hugh of Saint Victor, *Didascalicon* 1.5 (p. 52), 1.1 (p. 47). "To restore the divine likeness," 2.1 (p. 61).

36. It was this mental reordering that was the key, rather than the mastery of the mechanical arts as Caroline Merchant has suggested. *Reinventing Eden: The Fate of Nature in Western Culture* (London: Routledge, 2004), pp. 56 f. Hugh, for example, clearly regards the arts as intellectual exercises that provide access to the divine ideas that lie behind the material creation.

37. For Calvin's teachings on utility, see Calvin, *Harmony of the Gospels*, Matt. 25:15, *Calvin's Commentaries* XVII, 443. Cf. *Calvin's Commentaries* VI, 104; XXI, 115. For the impact of his ideas in seventeenth-century England, see David Little, *Religion, Order, and Law: A Study in Pre-Revolutionary England* (New York: Harper and Row, 1969), esp. p. 60.

38. Charlotte Methuen, "Interpreting the Books of Nature and Scripture in Medieval and Early Modern Thought: An Introductory Essay," in van der Meer and Mandelbrote, *Nature and Scripture*, vol. 1, 179–218; de Lubac, *Medieval Exegesis*, vol. 1, p. 51; Smalley, *The Study of the Bible in the Middle Ages*, 3rd ed. (Oxford: Blackwell, 1983), viii.

39. "*Nihil est in intellectu, quod non antea fuerit in sensu.*" Aquinas, ST 1a, 1, 9.

40. Aquinas, ST 1a, 13, 5. Cf. "in this life we cannot see the essence of God; but we know God from creatures as their principle, and also by way of excellence and remotion," ST 1a, 13, 1. Cf. "the creatures are the representations of God, although in an imperfect manner," ST 1a, 13, 2; "our knowledge of God is derived from the perfections which flow from Him to creatures. . . . Now our intellect apprehends them as they are in creatures," ST 1a, 13, 3.

41. Aquinas, ST 1a, 2, 2.

42. See, e.g., Aquinas, SCG 3.20–21.

43. Albertus Magnus, *Quaestiones super de animalibus* 9.3 (ET FC pp. 351 f.). Albert points out that this is knowledge "that it is" (*quia est*) not "what it is" (*quid est*) — knowledge through effects, not knowledge of the essence. Cf. Aquinas: "For wisdom is twofold: mundane wisdom called philosophy, which considers the lower causes, causes namely that are themselves caused, and bases its judgements on them: and divine wisdom or theology, which considers the higher, that is the divine, causes and judges according to them." *On the Power of God* Bk. 1, Q. 1, A. 4, Body (London, 1932), p. 24.

44. Aquinas SCG 3.70.108 (vol. 3/1, pp. 235–37).

45. William of Auvergne, *De universo* 1a.2ae, Preface, *The Universe of Creatures*, trans. Roland J. Teske (Milwaukee: Marquette University Press, 1998), p. 139.

46. Albertus Magus, *De bono* 4 Q. 1, art. 2, sol. (4), in *Opera Omnia* (Münster: Aschendorff, 1968), vol. 28 (1951), 224, quoted in Brague, *Legend of the Middle Ages*, p. 79. Cf. Brague, *Wisdom of the World*, pp. 118 f.

47. Aquinas ST 2a2ae, 180, 4.

48. de Lubac, *Medieval Exegesis*, vol. 1, p. 27.

49. Aquinas, ST 1a, 1, 3.

50. John I. Jenkins, *Knowledge and Faith in Aquinas* (Cambridge: Cambridge University Press, 1997), p. 5.

51. Thus Bonaventure: "Theological science is an affective habit and the mean between the speculative and practical, and for (its) end it has both contemplation, and that we become good, and indeed more principally, that we become good." *Commentaria in Quatuor Libros Sententiarum, Opera Omnia S. Bonaventurae* (Quaracchi: Ad Claras Aquas, 1882–1902), I:13. Also see the discussion in Aquinas, ST 2a2ae, 8, 6; SCG 3.26.1.

52. "To say that a man discharges his proper office is equivalent to saying that he acts virtuously." Aquinas, *Compendium of Theology* 1.1.172, trans. C. Vollert (Saint Louis: Herder, 1948), pp. 186 f. The passage from Aristotle cited in support of this definition is *Nicomachean Ethics* 1106a15.

53. Aquinas, ST 1a, 12, 5. More specifically, "when any created intellect sees the essence of God, the essence of God itself becomes the intelligible form of the intellect." Ibid. Cf. SCG 3.19. Thomas's emphasis on deification is owing partly to the influence of the neo-Platonized Aristotelianism found in medieval Arab sources. See Fergus Kerr, *After Aquinas: Visions of Thomism* (Oxford: Blackwell, 2002). This notion of an inner transformation was reinforced by Thomas's understanding of the transformative power of the mass: "The difference between corporeal and spiritual food lies in this, that the former is changed into the substance of the person nourished . . . but spiritual food changes man into itself." ST 3a, 73, 3.

54. Aristotle, *Nicomachean Ethics* 1177b, 1178b, in *Complete Works of Aristotle*, 2 vols., ed. Jonathan Barnes (Princeton, 1984), vol. 2, pp. 1861, 1862.

55. Aquinas, ST 1a2ae, 113, 1. Cf. Robert Grosseteste, who spoke of man being raised to "a sharing of the form of God." *On the Six Days of Creation* 8.6.4, trans. C. Martin (Oxford: Oxford University Press, 1996), p. 232. On the history of the theme of personal transformation, see Gerhart B. Ladner, *The Idea of Reform: Its Impact on Christian*

Thought and Action in the Age of the Fathers (Cambridge, MA: Harvard University Press, 1959).

56. Jenkins, *Knowledge and Faith in Aquinas*, p. 49.

57. For Aquinas, the kind of knowledge of God at which human beings aim cannot by identified with innate knowledge of God, nor with demonstrative knowledge (science of God), nor even with knowledge gained through faith. For this reason, such knowledge is not possible in the present life. SCG 3.37–40.

58. William Alston, *Perceiving God: The Epistemology of Religious Experience* (Ithaca: Cornell University Press, 1991), p. 289. Compare this with Lord Gifford's understanding of the "science" of natural theology, which, as he expresses it in his bequest, is devoid of "reference to or reliance upon any supposed special exceptional or so-called miraculous revelation."

59. A search for *theologia naturalis* and variants in PL and Brepols's Library of Latin Texts yielded two hits: Augustine, *City of God* VII.6, and William of Occam, *Scriptum in librum primum Sententiarum* (ordinatio) op. theol., vol. 1, prologus, quaestio 12, p. 365, line 3.

60. Aquinas, ST 2a2ae, 94, 1. Cf. Augustine, *City of God* VI.5: "There are three kinds of theology, or reasoning concerning the gods: of these one is mythical, the other natural [*physicum*], the third civil"; Tertullian, *Ad nationes* 2.1: "the philosophers have ingeniously composed their physical [theology] (*physicum theologiae*) out of their own conjectures" [PL 1, 659]. The works of Marcus Terentius Varro (116–27 BC) survive only as fragments.

61. Bacon, *Novum Organum*, I, 96, *Works*, vol. 4, p. 93.

62. Hugh of Saint Victor, *De tribus deibus*, passim (PL 176, 811–38).

63. Bonaventure, *Breviloquium* II.12.

64. Ramon Sibiuda [Raymond of Sabunde], *Theologia naturalis seu liber creaturarum*, ed. F. Stegmüller (Stuttgart-Bad Cannstatt: Frommann, 1966), Prologus.

65. For a sustained argument that Aquinas is involved in natural theology precisely along the lines of Alton's definition, see Norman Kretzmann, *The Metaphysics of Creation: Aquinas's Natural Theology in Summa Contra Gentiles* (Oxford: Oxford University Press, 1998), esp. p. 7.

66. Kerr, *After Aquinas*, pp. 58–72; Eric L. Mascall, *He Who Is: A Study in Traditional Theism* (London: Darton, Longman and Todd, 1967), pp. 80–82; Edward Sillem, *Ways of Thinking about God: Thomas Aquinas and Some Recent Problems* (London: Darton, Longman and Todd, 1961); Thomas S. Hibbs, *Dialectic and Narrative in Aquinas: An Interpretation of the Summa Contra Gentiles* (Notre Dame, 1995). It is also significant that for Aquinas the assurance that we can know God from nature is itself given in divine revelation, and is not a postulate of reason.

67. Leo Elders, "Justification des 'cinq voies,'" *Revue Thomiste* 61 (1961): 207–25. The five ways can be mapped onto the "three ways" of ascent to God set out by Dionysius (*via negationis, via causalitatis, via eminentiae*). For Elders, admittedly, they are not to be understood *only* in this sense. See *The Philosophical Theology of Aquinas* (Leiden: Brill, 1990).

68. Aquinas, SCG I.4.3 (vol. 1, p. 67).

69. Aquinas's distinction in SCG 1.3.2 is between two modes of truth—those accessible to natural reason, and those that exceed the ability of human reason.

70. *Nicholas of Lyra: The Sense of Scripture*, ed. Philip D. Krey and Leslie Smith (Leiden: Brill, 2000), pp. 1–18; Harrison, *Bible and the Rise of Natural Science*, pp. 107–20.

71. Martin Luther, *The Babylonian Captivity of the Church*, in *Three Treatises* (Philadelphia: Fortress Press, 1970), pp. 146, 241; cf. Luther, *Answer to the Hyperchristian Book*, in *Luther's Works*, 55 vols., ed. J. Pelikan and H. Lehman (Saint Louis: Concordia, 1955–75), 39, 177. These statements notwithstanding, the Reformers offered moral and typological readings of scripture, but these were not allegorical.

72. Ian Hazlett, "Calvin's Latin Preface of His Proposed French Edition of Chrysostom's Homilies: Translation and Commentary," in *Humanism and Reform*, ed. James Kirk (Oxford: Oxford University Press, 1991), pp. 129–50; John Calvin, *The Epistle of Paul the Apostle to the Galatians, Philippians, Ephesians, and Colossians*, trans. T. H. L. Parker (Grand Rapids: Eerdmans, 1964), pp. 84 f.

73. Bacon, *Advancement of Learning*, *Works*, vol. 3, pp. 349 f.

74. Plato, *Timaeus*, 92c. This is repeated in the hermetic literature. *Asclepius* 8 (Copenhaver ed. p. 71); cf. *Corpus Hermeticum*: the cosmos is "a great god and image of a greater"; XII.15 (Copenhaver ed. p. 46).

75. Origen, *The Song of Songs, Commentary and Homilies* (ACW vol. 26, pp. 218 f., and passim); Bonaventure, *Breviloquium* II.12.

76. Heinrich Cornelius Agrippa (1486–1535) thus contended that "the world is the image of God," a view that he attributed to Hermes Trismegistus. *Three Books of Occult Philosophy* (London, 1641), [ch. 36] pp. 457 f. Cf. also John Edwards, *A Demonstration of the Existence and Providence of God* (London, 1696), p. 260.

77. Bacon, *Novum Organum*, I.xxiii, in *Works*, vol. 4, p. 51. Cf. Noah Biggs, *Mataeotechnia Medicinae Praxeos* (London, 1651), p. 33. Boyle's contemporary, the mathematician John Wallis, described Boyle's approach in similar Baconian terms: "you pursue nature as if by iron and fire . . . you follow to the most hidden secret recesses, and penetrate as if to its visceral parts." John Wallis, *Opera mathematica* (Oxford, 1699), vol. 1, p. 491.

78. Cf. Johannes Kepler, *Mysterium Cosmographicum*, trans. A. M. Duncan (Norwalk, CT, 1999), pp. 53 f. Kepler explicitly opposed his cosmology to that of the "pagan" Aristotle, partly on the grounds that the latter had not believed in a divine creator. Ibid., p. 125, n. 2. Also *The Harmony of the World*, trans. and introduced by E. J. Aiton, A. M. Duncan, and J. V. Field (Philadelphia: American Philosophical Society, 1997), p. 115. In *Mysterium Cosmographicum*, Kepler does say that the earth "was to provide and nourish a true image of the Creator" (p. 107), but this is different from saying that it directly represents the Creator.

79. Galileo, *The Assayer*, in *Discoveries and Opinions*, trans. Stillman Drake, pp. 237–38. Cf. Galileo's *Dialogue concerning the Two Chief World Systems—Ptolemaic & Copernican*, trans. Stillman Drake (Berkeley: University of California Press, 1962), p. 3.

80. Descartes, *Principles of Philosophy* §61 CSM I, 24. For comparable statements, see

Robert Boyle, *The Christian Virtuoso*, in *Works*, 5:521; Samuel Clarke, *The Works of Samuel Clarke, D.D.* (London, 1738), 2 vols., vol. 2, p. 698. On the origins of the idea of laws of nature, see Walter Ott, *Causation and Laws of Nature in Early Modern Philosophy* (Oxford: Oxford University Press, 2009); John Henry, "Metaphysics and the Origins of Modern Science: Descartes and the Importance of Laws of Nature," *Early Science and Medicine* 9 (2004): 73–114; F. Steinle, "The Amalgamation of a Concept—Laws of Nature in the New Sciences," in *Laws of Nature: Essays on the Philosophical, Scientific and Historical Dimensions*, ed. F. Weinert (Berlin: De Gruyter, 1995), pp. 316–68; J. R. Milton, "Laws of Nature," in *The Cambridge History of Seventeenth Century Philosophy*, ed. D. Garber and M. Ayers (Cambridge: Cambridge University Press, 1998), 2 vols., vol. 1, pp. 680–701; Peter Harrison, "The Development of the Concept of Laws of Nature," in *Creation: Law and Probability*, ed. Fraser Watts (Aldershot: Ashgate, 2008), pp. 13–36; Peter Harrison, "Laws of Nature in Seventeenth Century England: From Cambridge Platonism to Newtonianism," in *God, Man and the Order of Nature*, ed. Eric Watkins (New York: Oxford University Press, 2014), pp. 127–48.

81. Descartes, *Principles of Philosophy* in CSM vol. 1, p. 256. Cf. *Isaac Newton: The Principia*, trans. I. B. Cohen and A. Whitman (Berkeley: University of California Press, 1999), p. 393 (preface by Roger Cotes).

82. Robert Boyle, *Usefulness of Natural Philosophy*, Works, II, pp. 20, 51–52, 62–63. For the idea of dissection as a kind of hermeneutical technique, I am indebted to Karen L. Edwards, *Milton and the Natural World* (Cambridge: Cambridge University Press, 1999), pp. 59–60. Boyle also penned an earlier unpublished work on the metaphor. See *Of the Study of the Book of Nature*, in *The Works of Robert Boyle*, 14 vols., ed. Michael Hunter and Edward Davis (London, 2000), vol. 13.

83. "The story of every beast is amplified with narrations out of scriptures, fathers, phylosophers, physicians, and poets: wherein are declared divers hyeroglyphicks, emblems, epigrams, and other good histories." Edward Topsel, *The Historie of Foure-Footed Beastes* (London, 1607), title page.

84. Nehemiah Grew, *Musaeum Regalis Societatis* (London, 1681), Preface.

85. John Ray and Francis Willoughby, *The Ornithology of Francis Willughby* (London, 1678); cf. *The Wisdom of God Manifested in the Works of Creation* (London, 1691), p. 124.

86. For a comprehensive account of the demise of Aristotelian understandings of causation, see Vincent Carraud, *Causa sive Ratio: La Raison de la cause, de Suarez à Leibniz* (Paris: Presses Universitaires de France, 2002).

87. An alternative source of internal causal order were the "seminal reasons" (*rationes seminales*) that, according to Augustine, God had implanted in natural things. This originally Stoic idea is also found in Plotinus and a number of medieval and early modern thinkers. See, e.g., Marcia L. Colish, *The Stoic Tradition from Antiquity to the Early Middle Ages* (Leiden: Brill, 1985), vol. 2, pp. 204–6; Jules Brady, "Saint Augustine's Theory of Seminal Reasons," *New Scholasticism* 38 (1964): 141–58; Peter Anstey, "Boyle on Seminal Priciples," *Studies in History and Philosophy of Science, Part C*, 33 (2002): 597–630. While seminal reasons proved attractive to some early modern philosophers, this conception was also eventually to succumb to the notion of laws of nature.

88. Aquinas, SCG 3.66, 3.77. Francisco Suárez, *On Efficient Causation*, trans. Alfred Freddoso (New Haven: Yale University Press, 1994), 18.1 (p. 14). According to one view—"conservationism"—God, in his ongoing creative act, conserves the active powers of things that operate more or less autonomously. On the more common view—"concurrentism"—God acts through the powers of things. On this latter account, espoused by both Aquinas and Suárez, a natural event is wholly the effect of God and the natural agent. See Aquinas, SCG 3.70.5. On creationism and concurrentism, see Alfred Freddoso, "God's General Concurrence with Secondary Causes: Why Conservation Is Not Enough," *Philosophical Perspectives* 5 (1991): 553–85.

89. Nicholas of Autrecourt (1300–ca. 1350) and Gabriel Biel (ca. 1425–95) were skeptical about Aristotelian causation. See Alfred Freddoso, "Medieval Aristotelianism and the Case against Secondary Causation in Nature," in *Divine and Human Action: Essays in the Metaphysics of Theism*, ed. Thomas V. Morris (Ithaca: Cornell University Press, 1988). Nicholas's arguments seemed to have drawn upon the philosophy of the great Islamic philosopher and theologian al-Ghazālī (ca. 1055–1111), who understood causality, as Hume was later to do, in terms of a human tendency to assume regularity on the basis of habitual experience. See Abū Hāmid al-Ghazālī, *The Incoherence of the Philosophers* (*Tahâfut al-falâsifa*), trans. P. Marmura (Salt Lake City: Brigham Young University Press, 2002), p. 170; Harry Wolfson, "Nicolaus of Autrecourt and Ghazali's Argument against Causality," *Speculum* 44 (1969): 234–38.

90. More recently, some philosophers have sought to resurrect the Aristotelian notion of laws of nature as descriptive of relations between the properties or dispositions of objects. For a concise account of the range of contemporary positions on laws of nature, see Ott, *Causation and Laws of Nature*, p. 7, n. 7.

91. Arguably, Hume replaced efficient causation (an ontological category) with one of its empirical tests—constant conjunction (an epistemic category). Mario Bunge, *Causality and Modern Science* (New York: Dover, 1979), p. 327.

92. William Whiston, *A New Theory of the Earth* (London, 1696), pp. 6, 211; Samuel Clarke, *Truth and Certainty of the Christian Revelation*, in *The Works of Samuel Clarke, D.D.* (London, 1738), 2 vols., vol. 2, p. 697; Richard Bentley, *The Works of Richard Bentley, D.D.*, ed. Alexander Dyce (London: Macpherson, 1838), vol. 3, pp. 74, 75. Also Harrison, "Laws of Nature in Seventeenth-Century England."

93. Samuel Clarke, *Works*, vol. 2, pp. 297–98.

94. For an account of the significance of Scotus's assertion of the "univocity of being," see Brad Gregory, *The Unintended Reformation* (Cambridge, MA: Harvard University Press, 2012), pp. 25–73. For the thesis that this was crucial to the rise of secular modernity, see, e.g., Louis Dupré, *Passage to Modernity* (New Haven: Yale University Press, 1993); Michael Gillespie, *The Theological Origins of Modernity* (Chicago: University of Chicago Press, 2008); John Milbank, *Theology and Social Theory: Beyond Secular Reason*, 2nd ed. (Oxford: Wiley-Blackwell, 2006), pp. 302–6. But cf. Richard Cross, "Duns Scotus and Suárez at the Origins of Modernity," in *Deconstructing Radical Orthodoxy: Postmodern Theology, Rhetoric and Truth*, ed. Wayne J. Hankey and Douglas Hedley (Aldershot: Ashgate, 2005), pp. 65–80; Michael J. Dodds, *Unlocking Divine Action:*

Contemporary Science and Thomas Aquinas (Washington, DC: Catholic University of America Press, 2012), p. 52; Harrison, "'Voluntarism and Early Modern Science': Voluntarism and the Origins of Modern Science: A Reply to John Henry," *History of Science* 47 (2009): 223–31; Taylor, *A Secular Age*, pp. 773–74.

Chapter Four

1. James Boswell, *Life of Johnson*, 4 vols., ed. George B. Hill and Lawrence Powell (Oxford: Clarendon, 1994), vol. 1, p. 398.

2. See, e.g., Scott Atran, *In Gods We Trust: The Evolutionary Landscape of Religion* (Oxford: Oxford University Press, 2002); Pascal Boyer, *Religion Explained: The Evolutionary Origins of Religious Thought* (New York: Basic Books, 2001); *The Naturalness of Religious Ideas: A Cognitive Theory of Religion* (Berkeley: University of California Press, 1994); Daniel Dennett, *Breaking the Spell: Religion as a Natural Phenomenon* (Harmondsworth: Penguin, 2006); Robert Hinde, *Why Gods Persist: A Scientific Approach to Religion* (London: Routledge, 1999).

3. David Sloan Wilson, *Darwin's Cathedral: Evolution, Religion and the Nature of Society* (Chicago: University of Chicago Press, 2002), p. 93.

4. Ibid., p. 1.

5. Alasdair MacIntyre, *After Virtue*, 2nd ed. (Notre Dame: University of Notre Dame Press, 1984), esp. pp. 51–61.

6. Luther, "Sunday after Christmas, 6," in *The Complete Sermons of Martin Luther*, 7 vols., ed. John N. Lenker (Grand Rapids: Baker Books, 2000), vol. 3, p. 226. Cf. Aristotle, *Nichomachean Ethics* 1106a15; Aquinas, ST 2a2ae, 58, 3; ST 1a2ae, 55, 3 and 4; ST 2a2ae, 58, 3.

7. Luther, *D. Martin Luthers Werke* (Weimar, 1883–1948), 39^r, 278, trans. in Paul Althaus, *The Theology of Martin Luther* (Philadelphia: Fortress, 1966), p. 156, n. 71. The analogy of the harp player is Aristotle's from book 2 of the *Nicomachean Ethics*. For Aquinas's explicit appropriation of the Aristotelian idea of habit, see Aquinas, *Commentary on the Nicomachean Ethics* I, Bk. 1, Lec. 20, §244 (p. 105). For Luther's criticism of the Aristotelian notion of *habitus*, see Gerhard Ebeling, *Luther: An Introduction to His Thought* (London: Collins, 1970), pp. 150–58; Peter Nickl, *Ordnung der Gefühle: Studien zum Begriff des habitus* (Hamburg: Felix Meiner Verlag, 2001), pp. 118 f. Terence Irwin, "Luther's Attack on Self-Love: The Failure of Pagan Virtue," *Journal of Medieval and Early Modern Studies* 42 (2012): 151–55. For the medieval background, see Alister McGrath, *Iustitia Dei: A History of the Christian Doctrine of Justification*, 3rd ed. (Cambridge: Cambridge University Press, 2005), pp. 55–207. See also Gregory, *Unintended Reformation*, pp. 207–11.

8. Luther, *Lectures on Galations, Luther's Works*, ed. J. Pelikan and H. Lehman (Saint Louis: Concordia, 1955–75), 26, p. 234. Cf. *Lectures on Romans, Luther's Works*, vol. 25, p. 162.

9. Calvin, *Tracts and Treatises on the Reformation of the Church*, 7 vols., trans. Henry Beveridge (Edinburgh, 1844), vol. 3, p. 247. Cf. *Institutes of the Christian Religion* III.xv; James Sadolet, "Letter to the Senate and People of Geneva," in Calvin, *Tracts and*

Treatises, vol. 1, p. 9; and *Tracts and Treatises*, vol. 3, pp. 117, 153. Calvin's reference to "infused" virtues alludes to the fact that the Catholic position had been that the supernatural virtues are infused by God rather than being acquired through practice. The question then became what occasioned the infusion of the supernatural virtues, and what sense could be made of the notion of a "divinely infused habit."

10. David C. Steinmetz, "What Luther Got Wrong," *Christian Century*, 23 August 2005, 23–25 (23).

11. The Jesuit Benedict Pereira, e.g., taught that "the primary end of learning is the recognition of truth, which is the perfection of the human mind," in Ladislaus Lukács, S.J., *Monumenta Paedagogica Societatis Iesu* (Rome: Institutum Historicum Societatis Iesu, 1974), vol. 2, p. 670.

12. "Ignorant of themselves, of God much more, / And how the World began, and how Man fell, / Degraded by himself, on grace depending? / Much of the Soul they talk, but all awry; / And in themselves seek virtue; and to themselves / All glory arrogate, to God give none"; Milton, *Paradise Regained*, Bk. 4, lines 310–15.

13. Calvin, *Commentary on Genesis*, 3:6, *Calvin's Commentaries*, vol. 1, p. 154. Cf. Duns Scotus, *In sententias* 2.29.1, *Opera omnia* vol. 13, pp. 267 f.

14. Aristotle, *Metaphysics* 980a21; *Nicomachean Ethics* 1177a.

15. John Duns Scotus, *De cognitione humana*, in *Duns Scotus: Philosophical Writings*, trans. Allan Wolter (Indianapolis: Hackett, 1987), pp. 99–100. See also Aquinas, SCG 3.25.12. To some degree, this notion was also implicit in a common medieval understanding of truth as *adaequatio intellectus et rei*, which is to say, as a correspondence between intellect and thing. See, e.g., Aquinas, *Quaestiones disputatae de veritate* QI, Art. I.

16. I Cor. 13:12. The Greek ἐσόπτρου (*esoptrou*) is often translated "mirror" or "lens." Bacon, *The Great Instauration*, in *Works*, vol. 4, pp. 27, 7.

17. Robert Hooke, *Micrographia* (London, 1665), Preface.

18. For the manner in which the idea of original sin informed experimental natural philosophy, see Peter Harrison, *The Fall of Man and the Foundations of Science* (Cambridge: Cambridge University Press, 2007).

19. Whether the voluntaristic character of these divine laws was crucial in these developments has been a matter of some discussion. From an extensive literature, see M. B. Foster, "The Christian Doctrine of Creation and the Rise of Modern Natural Science," *Mind*, n.s., 43 (1934): 446–68; Francis Oakley, "Christian Theology and the Newtonian Science: The Rise of the Concept of Laws of Nature," *Church History* 30 (1961): 433–57; Margaret J. Osler, *Divine Will and the Mechanical Philosophy: Gassendi and Descartes on Contingency and Necessity in the Created World* (Cambridge: Cambridge University Press, 1994); John Henry, "Voluntarist Theology at the Origins of Modern Science: A Response to Peter Harrison," *History of Science* 47 (2009): 79–112. For my reservations about the voluntarism and science thesis, see Peter Harrison, "Voluntarism and Early Modern Science," *History of Science* 40 (2002): 63–89; "Was Newton a Voluntarist?," in *Newton and Newtonianism: New Studies*, ed. James E. Force and Sarah Hutton (Dor-

drecht: Kluwer, 2004), pp. 39–64; "Voluntarism and the Origins of Modern Science: A Reply to John Henry," *History of Science* 47 (2009): 223–31.

20. Samuel Butler, *Characters and Passages from Notebooks,* ed. A. R. Waller (Cambridge: Cambridge University Press, 1908), p. 281.

21. Francis Bacon, *New Organon* pt. 2, §2: "the final cause rather corrupts than advances the sciences." *Works,* vol. 4, p. 120. Descartes, *Meditations* in CSM vol. 2, p. 38, cf. p. 258. Descartes does concede that the use of final causes is more permissible in ethics, since speculation is more legitimate in that sphere. For nuances of the Cartesian position, see Allison J. Simmons, "Sensible Ends: Latent Teleology in Descartes' Account of Sensation," *Journal of the History of Philosophy* 39 (2001): 49–75. See also Thomas Hobbes, *Elements of Philosophy* II.10.7, in *The English Works of Thomas Hobbes,* ed. William Molesworth (London: Bohn, 1839), vol. 1, pp. 131–32.

22. "The Westminster Larger Catechism," Q. 1. That the chief end of man was the glorification and enjoyment of God was established in I Cor. 10:31; Rom. 11:3; Ps. 73:24–26; John 17:22, 24.

23. Thus Puritan divine William Ames: "But if those imperfect notions concerning that which is honest, and dishonest, be understood, which are found in the mind of man after the fall: seeing they are imperfect and very obscure, they cannot exactly informe vertue; neither indeed doe they differ any thing from the written Law of God, but in imperfection and obscurity only." *The Marrow of Sacred Divinity* (London, 1642), p. 226. Ames goes on to critique the Thomist typology of gifts of the spirit, fruits of the spirit, and beatitudes (pp. 228–29).

24. John Flavell, *An Exposition of the Assembly's Shorter Catechism* (London, 1688). Man's secondary end was set out in Gen. 1:26 and Ps. 8:6.

25. See esp. Corneanu, *Regimens of the Mind.*

26. John Locke, *Essay concerning Human Understanding* [1690] 2.28.11, 2.28.14, ed. Peter H. Nidditch (Oxford: Clarendon, 1975), pp. 356, 357. Locke describes three sources of moral law—"first, the law of God; secondly, the law of politic societies; thirdly, the law of fashion or private censure." Ibid. 2.28.13 (p. 357). A similar view of virtue is also manifest in Kant's *Metaphysics of Morals;* see Jerome B. Schneewind, "The Misfortunes of Virtue," in *Virtue Ethics,* ed. Roger Crisp and Michael Slote (Oxford: Oxford University Press, 1997), pp. 178–200. Even those who argued for the continuing value of Aristotelian moral philosophy sought to rearticulate it in such a way that it conformed to the priority given to divine commandments. See M. W. F. Stone, "The Adoption and Rejection of Aristotelian Moral Philosophy in Reformed 'Casuistry,'" in *Humanism and Early Modern Philosophy,* ed. Jill Kraye and M. W. F. Stone (London: Routledge, 2000), pp. 59–90.

27. John McDowell, "Virtue and Reason," in Crisp and Slote, eds., *Virtue Ethics,* p. 141.

28. For Descartes's rejection of the idea of *habitus,* see Jean-Luc Marion, *L'Ontologie grise de Descartes* (Paris: Vrin, 1971), pp. 25–30; Nickl, *Ordnung der Gefühle,* pp. 133–42. Part of what was at issue was whether knowledge of the multiplicity of natural things,

distinguished by their substantial forms, called for a multiplicity of distinct mental habits. Descartes sought to replace multiple habits with a single *mathesis universalis*, consistent with his reduction of the material realm to particles of matter in motion. *Rules for the Direction of the Mind*, CSM vol. 1, pp. 9, 19.

29. See the discussion in Han van Ruler, *The Crisis of Causality: Voetius and Descartes on God, Nature, and Change* (Leiden: Brill, 1995), pp. 305–19. The translated quotation from Voetius is given on p. 309.

30. See, e.g., Aquinas, ST 2a2ae, 2, 6.

31. Calvin, *Institutes* (1536), Prefatory Address (Beveridge ed., vol. 1, p. 7). The Latin version begins: *"Veram religionem, quae Scripturis tradita est, quaeque inter omnes constare debuerat."* Classical Latin has no articles, although in Medieval Latin *hic* and *ille* might be used to give the same sense as the definite article. In this passage, English translators inserted the definite article, and this is not inconsistent with the meaning of the rest of the sentence.

32. John Calvin, *The Institution of Christian Religion*, trans. Thomas Norton (London, 1651), Sig. A iii r. This usage was mitigated somewhat by the fact the title omits the definite article—*The Institution of Christian Religion*. Calvin, incidentally, was well aware of the classical understanding of *religio* as service owed to the gods. See *Calvin's Commentary on Seneca's De Clementia*, ed. Ford Lewis Battles (Leiden: Brill, 1969), pp. 363–65.

33. "The Book of Concord," http://bookofconcord.org/lc-2-preface.php, accessed 16 April 2010.

34. Ian Green, *The Christian's ABC: Catechisms and Catechizing in England c. 1530–1740* (Oxford: Clarendon, 1996), p. 45.

35. William Perkins, *A Golden Chaine* (London, 1600), Epistle Dedicatory.

36. Jean Delumeau, *Catholicism between Luther and Voltaire* (London: Burns and Oates, 1977). Cf. Charles Taylor, *A Secular Age* (Cambridge, MA: Harvard University Press, 2007), p. 774.

37. *"Scientia duobus modis accipitur: proprie pro eo habitus, quem per demonstrationen acquirimus. . . . Scientia, inquit Themistius, est notitia per demonstrationem acquisita."* *"Scientia accipitur, proprie pro notiti demonstrate acquisita. Improprie pro quibusuis habitibus intellectiuis."* Rudolph Goclenius, *Lexicon Philosophicum* (Frankfurt, 1613), pp. 1009, 1012. Cf. the entry for *"Habitus,"* p. 624. I am grateful to Dan Garber for this reference.

38. Eustachius à Sancto Paulo, *Summa Philosophiae Quadripartita* (Paris, 1609), I, 230–31; in Étienne Gilson, *Index Scholastico-Cartésien* (Paris, 1912; reprinted New York: Franklin, 1964), p. 262. I am grateful to Stephen Gaukroger for drawing this passage to my attention. Robert Recorde, *The Urinal of Physick* (London, 1651), p. 137. Robert South, *Twelve sermons upon several subjects and occasions*, vol. 3 (London, 1698), p. 75. See also Francisco Toledo, *Commentaria una cum quæstionibus in universam Aristotelis logicam Post. anal.*, lib. II, cap. 18, qu. 1 (Rome, 1572), in Gilson, *Index Scholastico-Cartésien*, p. 54. William Vaughan, *The golden-groue moralized in three bookes: a worke very necessary for all such, as would know how to gouerne themselues, their houses, or their countrey* (London, 1600), Bk. 1, ch. 65 (unpaginated).

39. Taylor, *A Secular Age*, pp. 90–145. Also Philip S. Gorski, *The Disciplinary Revolution: Calvinism and the Rise of the State in Early Modern Europe* (Chicago: University of Chicago Press, 2003); Gregory, *Unintended Reformation*, pp. 209–10.

40. Michel Foucault, *Discipline and Punish* (Vintage, 1979), p. 141 and passim.

41. Bacon, *Advancement of Learning, Works*, vol. 3, p. 294.

42. Bacon, *Valerius Terminus, Works*, vol. 3, p. 222. Cf. "For the matter in hand is no mere felicity of speculation, but the real business and fortunes of the human race, and all power of operation." *Great Instauration, Works*, vol. 4, p. 32. For a convincing argument to the effect that Bacon introduced a new conception of natural philosophy, see Stephen Gaukroger, *Francis Bacon and the Transformation of Early Modern Natural Philosophy* (Cambridge: Cambridge University Press, 2001), p. 5. Cf. Antonio Pérez-Ramos, "Bacon's Legacy," *Cambridge Companion to Bacon*, pp. 311–34.

43. All of the developments described in these few pages are part of what historians now routinely refer to as the confessionalization of Europe. Importantly, the processes of confessionalization were not restricted to Protestant territories (although arguably confessionalization was more effective there). For the link between the disciplinary society and confessionalization, see R. W. Scribner, *Popular Culture and Popular Movements in Reformation Germany* (London: Hambledon, 1987), esp. pp. 175–84. On the confessionalization thesis generally, see Heinz Schilling, *Konfessionskonflikt und Staatsbildung* (Gutersloh, 1981); Stefan Ehrenpreis and Ute Lotz-Heumann, *Reformation und konfessionelles Zeitalter* (Darmstadt: Wissenschaftliche Buchgesellschaft, 2002); Heinrich Richard Schmidt, *Konfessionalisierung im 16. Jahrhundert* (Munich: Oldenbourg, 1992); Harm Klueting, *Das konfessionelle Zeitalter, 1525–1648* (Stuttgart: E. Ulmer, 1989). For a sociological perspective, see Peter Beyer, "The Modern Emergence of Religions and a Global Social System for Religion," *International Sociology* 13 (1998): 151–72. Arguably, a model for such territorialization of religion was provided by the expulsion of Jews and Muslims from the Iberian Peninsula by Catholic monarchs in the late fifteenth century.

44. "The Religious Peace of Augsburg (1555)," Article 17, in *Documents from the History of Lutheranism, 1517–1750*, ed. Eric Lund (Minneapolis: Augsburg Fortress, 2002), p. 170.

45. On this theme, see William T. Cavanaugh, *The Myth of Religious Violence: Secular Ideology and the Roots of Modern Conflict* (Oxford: Oxford University Press, 2009); Craig Calhoun, Mark Juergensmeyer, and Jonathan VanAntwerpen, eds., *Rethinking Secularism* (Oxford: Oxford University Press, 2011), introduction, pp. 14–20; Richard King, "The Association of Religion with Violence: Reflections on a Modern Trope," in *Religion and Violence in South Asia*, ed. John Hinnells and Richard King (Abingdon: Routledge, 2007), pp. 214–42.

46. Samuel Preus has thus suggested that Puritans, *because of* their deeply felt religious convictions, were among the most prominent reifiers of religion. J. Samuel Preus, "The Reified Heart in Seventeenth-Century Religion," in *Religion in History: The Word, the Idea, the Reality*, ed. Michel Despland and Gérard Vallée (Waterloo: Wilfrid Laurier University Press, 1992), pp. 45–56.

47. Christian Thomasius, "On the right of Protestant princes regarding heretics" (1697), §§39–41, in Ian Hunter, *The Secularization of the Confessional State: The Political Thought of Christian Thomasius* (Cambridge: Cambridge University Press, 2007), pp. 168–206 (181). See also Martin Heckel, "Das Säkularisierungsproblem in der Entwicklung des deutschen Staatskirchenrechts," in *Christentum und modernes Recht; Beiträger zum Problem der Säkularisation*, ed. G. Dilcher and I. Staff (Frankfurt am Main: Suhrkamp, 1984), pp. 35–95. I am grateful to Ian Hunter for these references. In certain respects, Thomasius represents a remarkable anticipation of Wilfred Cantwell Smith, who also sees the modern problem of the competing truth claims of the religions as an artifact of the category "religion."

48. Harrison, *"Religion" and the Religions*, pp. 9–10; Philip C. Almond, *The British Discovery of Buddhism* (Cambridge: Cambridge University Press, 1988); Tomoko Masuzawa, *The Invention of World Religions* (Chicago: University of Chicago Press, 2005); Arvind Mandair, *Religion and the Specter of the West: Sikhism, India, Postcoloniality, and the Politics of Translation* (New York: Columbia University Press, 2009); Brent Nongbri, *Before Religion: The History of a Modern Concept* (New Haven: Yale University Press, 2012).

49. Jean Bodin, *Colloquium of the Seven about the Secrets of the Sublime*, trans. Marion Kurtz (Princeton: Princeton University Press, 1975), p. 256.

50. Robert Boyle, "The Diversity of Religions," in *The Works of Robert Boyle*, ed. Michael Hunter and Edward B. Davis, 12 vols. (London: Chatto and Pickering, 1999–2000), vol. 14, p. 264 (Boyle Papers, vol. 4, fol. 281).

51. Thus, for example, "How to find out the true Faith & Religion it is a matter of very great difficulty . . . by reason that there are many faiths and religions in the world, and of all these there is but one true, and all the rest be false." This statement appears in the preface to a seventeenth-century translation of Augustine's *De utilitate credendi*. The translator, A.P., gives the work this extended title: *The Profit of Believing: Very usefull Both for all those that are not yet resolved what Religion they ought to embrace: And for them that desire to know whither their Religion be true or no* (London, 1651). Another influential writer on this topic was Edward, Lord Herbert of Cherbury, *De Religione Laici* [1645], ed. and trans. Harold R. Hutcheson (New Haven: Yale University Press, 1944) (see esp. p. 87). See also Samuel Pufendorf, *Nature and Qualification of Religion in Reference to Civil Society*, trans. Jodocus Crull, ed. Simone Zurbuchen (Indianapolis: Liberty Fund, 2002), p. 11.

52. Anthony Gilby and William Kethe, *The appellation of Iohn Knoxe . . .* (Geneva, 1558), p. 12v. Cf. Richard Bernard: "The Christian Religion taught at Ierusalem by the Apostles, and other Disciples of Christ, was our Religion, as by the first Argument is prooued from the Apostles writings." *Look beyond Luther* (London, 1623), p. 28. Also Heinrich Bullinger, *Fiftie Godly and Learned Sermons* (London, 1577), p. 181; Nicholas Byfield, *The Rule of Faith* (London, 1626), p. 307; John Owen, *Animadversions on a Treatise Entitled "Fiat Lux,"* in *Works of John Owen, D. D.*, 16 vols. (Edinburgh: T. & T. Clark, 1842), vol. 14, pp. 19 f.

53. Jeremy Taylor, *Symbolon theologikon* (London, 1674), p. 6.

54. John Owen, *Truth and Innocence Vindicated* (London, 1669), p. 30.

55. Thomas Sprat, *Sermons Preached upon Several Occasions* (London, 1697), p. 2.

56. For examples, see Harrison, *"Religion" and the Religions*, pp. 19–28.

57. Robert Boyle, Boyle Papers at the Library of the Royal Society of London, vol. 4, fol. 74r; vol. 1, fol. 61r.

58. Although these "evidences" were not then necessarily understood in terms of what we would regard as objective proofs, they were increasingly so from the beginning of the seventeenth century. For many Puritan writers, however, "experimental" evidence of religious convictions came from profoundly personal experiences. See Harrison, "Experimental Religion and Experimental Science in Early Modern England," *Intellectual History Review* 21 (2011): 413–33.

59. [Gabriel Plattes], *A Description of the Famous Kingdome of Macaria* (London, 1641). Authorship was traditionally ascribed to Samuel Hartlib, but Charles Webster has convincingly argued that it was the work of Gabriel Plattes. Charles Webster, "The Authorship and Significance of Macaria," *Past and Present* 56 (1972): 34–48.

60. John Henry Newman, *Fifteen Sermons Preached before the University of Oxford* (New York, 1918), p. 197. Prominent eighteenth-century works in the genre include Joseph Addison, *The Evidences of the Christian Religion* [1721] (London, 1807); William Paley, *View of the Evidences of Christianity*, 2 vols. (London, 1794); Soame Jenyns, *View of the Internal Evidence of the Christian Religion* (1776); Mark Hopkins, *Evidences of Christianity* [1846] (Boston, 1909).

61. Isaac Barrow, Sermon IV, "Of Justifying Faith," in *The Works of Isaac Barrow*, 3 vols. (New York: John Riker, 1845), vol. 2, p. 207.

62. Thomas Aquinas, ST 2a2ae, 1, 1. In the next question (1, 2) Aquinas nuances this by allowing that the object of faith may be considered in two ways, "namely the thing itself about which we have faith . . . and the object of faith as something complex by way of a proposition."

63. Thomas Aquinas, *De veritate* 14.8.9. Here lies the difference between faith, a theological virtue, and the intellectual virtues of science and understanding (ST 2a2ae, 1, 4). W. C. Smith has also argued that for Aquinas "assent" (*assensio, assensus*) means something more like "agreement, approbation, applause, approval." Smith, *Faith and Belief*, pp. 283 ff.; cf. Lewis and Short, *A Latin Dictionary* (Oxford: Clarendon, 1879), p. 177. This reading has its problems (see Frederick J. Crossin, "'Fides' and 'Credere': W. C. Smith on Aquinas," *Journal of Religion* 65 [1985]: 399–412), but it is plausible to suggest that "assent" for Aquinas retains some of its original links with the realm of feeling (*sentio, sensus*). What we can say is that in Aquinas the element of trust that is required by implicit faith extends beyond confidence in the utterances of ecclesiastical authorities to the source of revealed truths—God himself.

64. Aquinas thus upholds the Augustinian formula that distinguishes between believing God, believing in a God, and believing in/on God (*credere Deo, credere Deum, credere in Deum*). ST 2a2ae, 2, 2. Here Aquinas seems to imply that "God exists" means something different for Christians and non-Christians, because unbelievers "do not

truly believe in a God." See Victor Preller, *Divine Science and the Science of God: A Refor-mulation of Thomas Aquinas* (Princeton: Princeton University Press, 1967), p. 228. Cf. Augustine, *Homilies on the Gospel of John*, Tractate 29, 6; *Expositions of the Psalms* 77, 8. These distinctions relate to that other medieval distinction between *fides quae* and *fides qua creditur* (the faith that is believed, and the faith by which it is believed).

65. Thomas Aquinas, *Commentary on the Sentences of Peter Lombard* (Toronto: Pon-tifical Institute, 1997), pp. 24, 1, iii.

66. Francis Bacon, "Of Heresies," *Translation of the Meditationes Sacrae, Works*, vol. 14, p. 94. It is in this passage, incidentally, that Bacon makes his oft-misinterpreted remark that "knowledge is power" (*scientia potestas est*). The reference is to divine power, and in full the passage should be translated "his [God's] knowledge is his power."

67. Erasmus, *Ecclesiastes* [1535], in *Desiderii Erasmi Roterodami Opera Omnia*, ed. J. Clericus (Leyden, 1703–6), 10 vols., vol. 5, col. 1081B.

68. Locke, *Essay concerning Human Understanding*, pp. 44, 667. For Thomas, by way of contrast, it is the object of faith—God—who moves the will to give assent, and not reason and evidence. ST 2a2ae, 1, 4.

69. Locke, *Essay*, p. 99. For Locke's condemnation of implicit faith, see *A Third Let-ter for Toleration*, in *The Works of John Locke*, 12th ed. (1823) vol. 6, pp. 152, 407; Locke, *Of the Conduct of the Understanding*, 5th ed., ed. Thomas Fowler (Oxford: Oxford Univer-sity Press, 1901), p. 6; Locke: *A Second Vindication of the Reasonableness of Christianity*, in *Works of John Locke*, 12th ed. (1823), vol. 7, p. 296.

70. Locke, *Essay concerning Human Understanding*, p. 400. See also Locke, *A Third Letter for Toleration, Works*, vol. 6, pp. 203, 354, 457.

71. Locke, *A Third Letter for Toleration, Works*, vol. 6, pp. 63, 144. This identification is even stronger in the writing of Locke's correspondent, Jonas Proast, *The Argument of the "Letter concerning Toleration" considered and answered* (Oxford, 1690). Thomas Hobbes, the other great seventeenth-century English philosopher, had already expressed similar views about "faith," "belief in," and implicit faith in his *Leviathan* (1651). Noting that the expression "believe in" (*credo in*) occurs only in the writings of divines, he astutely observed that this usage "hath raised many disputes over the right object of the Christian faith." Hobbes concluded that "by *believing in*, as it is in the creed, is meant, not trust in the person; but confession and acknowledgement of the doctrine." *Leviathan* ch. 7, paras. 5–7, ed. C. B. Macpherson (Ringwood: Penguin, 1982), p. 130.

72. Further on this theme see Stephen Gaukroger, *The Emergence of a Scientific Cul-ture: Science and the Shaping of Modernity, 1210–1685* (Oxford: Oxford University Press, 2006), pp. 3, 7–8, 497; *The Collapse of Mechanism and the Rise of Sensibility: Science and the Shaping of Modernity, 1680–1760* (Oxford: Oxford University Press, 2010), p. 41; Peter Harrison, "Physico-Theology and the Mixed Sciences: The Role of Theology in Early Modern Natural Philosophy," in *The Science of Nature in the Seventeenth Century*, ed. Peter Anstey and John Schuster (Dordrecht: Springer, 2005), pp. 165–83. Aquinas had

explicitly distinguished the realm of faith from that of the intellectual virtues of science and understanding. ST 2a2ae, 1, 4. Cf. George Downame: "Science is begotten by virtue of demonstrative reason; so faith is not demonstrated but is undertaken by the virtue or power of the will." *A Treatise of Iustification* (London, 1633), p. 359.

73. Isaac Newton, *The Principia*, trans. I. Bernard Cohen and Anne Whitman (Berkeley: University of California Press, 1999), pp. 940, 943.

74. See, e.g., Andrew Cunningham, "How the *Principia* Got Its Name; or, Taking Natural Philosophy Seriously," *History of Science* 28 (1991): 377–92. But cf. Edward Grant, "God and Natural Philosophy: The Late Middle Ages and Sir Isaac Newton," *Early Science and Medicine* 6 (2000): 279–29.

75. Harrison, "Physico-Theology and the Mixed Sciences."

76. Amos Funkenstein, *Theology and the Scientific Imagination* (Princeton: Princeton University Press, 1986), pp. 4–10.

77. Johannes Kepler, *Gesammelte Werke* (Munich, 1937–45), XIII, 40. For Kepler's own account, see Kepler, *Selbstzeugnisse*, ed. Franz Hammer, trans. Esther Hammer (Stuttgart-Bad Constatt, 1971), pp. 61–65.

78. Johannes Kepler, *Mysterium Cosmographicum*, trans. A. M. Duncan (Norwalk, CT, 1999), p. 53.

79. Robert Boyle, *Some Considerations Touching the Usefulness of Experimental Natural Philosophy*, in *The Works of the Honourable Robert Boyle*, 6 vols., ed. Thomas Birch (Hildesheim, 1966) vol. 2, p. 31. For an early account of Boyle's notion of the priest-scientist, see H. Fisch, "The Scientist as Priest: A Note on Robert Boyle's Natural Theology," *Isis* 44 (1953): 252–65. See also Peter Harrison, "Sentiments of Devotion and Experimental Philosophy in Seventeenth-Century England," *Journal of Medieval and Early Modern Studies* 44 (2014): 113–33.

80. Boyle, *Usefulness of Natural Philosophy*, in *Works*, vol. 2, pp. 62 f.

81. Ibid., vol. 2, p. 32.

82. Hooke, *Micrographia*, p. 8. On Hooke and the argument from design, see John Harwood, "Rhetoric and Graphics in *Micrographia*," in *Robert Hooke: New Studies*, ed. Michael Hunter and Simon Schaffer (Woodbridge: Boydell, 1989), pp. 119–48.

83. Boyle, *Some Physico-Theological Considerations about the Possibility of the Resurrection* (London, 1675), Preface.

84. Boyle, *Usefulness of Experimental Natural Philosophy*, in *Works*, vol. 2, p. 20. Cf. p. 6.

85. Sprat, *History of the Royal Society*, pp. 349 f.

86. See, e.g., Richard Baxter, *Reasons of the Christian Religion* (1667), title page; Boyle, *Works* (Hunter and Davis ed.) vol. 14, p. 264.

87. Robert Boyle, *The Christian Virtuoso*, in *Works*, vol. 5, pp. 538, 358 f.

88. Sprat, *History of the Royal Society*, pp. 358 f.

89. Michael J. Buckley, *At the Origins of Modern Atheism* (New Haven: Yale University Press, 1987).

90. Hunter, *Secularization of the Confessional State*, p. 154.

91. John Wesley, "Salvation by Faith," Sermon Preached at Saint Mary's, Oxford, before the University, 18 June 1738, in *The Works of the Rev. John Wesley*, ed. John Emory, 3rd ed., 7 vols. (New York: Carlton and Porter, 1856), vol. 1, p. 14. On the persistence of "heart religion," see Phyllis Mack, *Heart Religion in the British Enlightenment* (Cambridge: Cambridge University Press, 2008); David Hempton, *Methodism: Empire of the Spirit* (New Haven: Yale University Press, 2005).

92. Karl Barth, *Church Dogmatics*, 4 vols. (Edinburgh: T. & T. Clark, 1936–69), vol. I/2, p. 288. Dietrich Bonhoeffer, *Letters and Papers from Prison* (New York: Macmillan, 1962), pp. 161–69, 194–200, 226. Also see examples in Smith, *Meaning and End of Religion*, pp. 125 f.

Chapter Five

1. Jonathan Swift, "Thoughts on Various Subjects, Moral and Diverting," in *The Works of the Rev. Jonathan Swift, D.D.*, 24 vols. (London, 1803), vol. 14, p. 165.

2. Thomas Babington Macaulay, *Critical, Historical and Miscellaneous Essays*, 6 vols. (New York: Hurd and Houghton, 1860), vol. 3, pp. 436, 447, 448, 463.

3. Macaulay, *The History of England from the Accession of James II*, 2 vols. (London: Longman, Brown, Green and Longmans, 1849), vol. 1, p. 14. On Whig history, see Herbert Butterfield, *The Whig Interpretation of History* (New York: W. W. Norton, 1965); Nick Jardine, "Whigs and Stories: Herbert Butterfield and the Historiography of Science," *History of Science* 41 (2003): 125–40.

4. Gertrude Lenzer, ed., *Auguste Comte and Positivism: The Essential Writings* (New York: Harper and Row, 1975), p. lxviii.

5. See, respectively, Charles Freeman, *The Closing of the Western Mind: The Rise of Faith and the Fall of Reason* (London: William Heinemann, 2002); A. C. Grayling, *Towards the Light: The Struggles for Liberty and Rights That Made the Modern West* (London: Bloomsbury, 2007); Francis Fukuyama, *The End of History and the Last Man* (New York: Free Press, 1992).

6. Aquinas, ST 1a2ae, 52, 2. Cf. ST 1a2ae, 54, 4; 2a2ae, 1, 7.

7. For typical examples, see ST 1a, 43, 6; 1a2ae, 69, 2; 1a2ae, 87, 7; 1a2ae, 114, 8; 2a2ae, 24, 9. Cf. Peter Lombard, *Sententiarum Quatuor Libri*, Bk. 2, Dist. 24, pt. 1, ch. 1 [*Opera Omnia S. Bonaventurae*, Ad Claras Aquas, 1885, vol. 2, pp. 549–53]. Peter Lombard speaks of *proficere* (advance) and *profectus* (progress).

8. Marcus Aurelius, V 16, trans. David Sedley, *Creation and Its Critics*, p. 237.

9. Pascal, "Preface to the Treatise on the Vacuum," in Pascal, *Thoughts, Letters, Minor Works* (New York: Cosimo Books, 2009), p. 449. It is significant that Pascal expounds this idea about cumulative learning in the context of his assertion of the existence of the vacuum—a possibility that had long been dismissed on the authority of the ancients. See also Bury, *The Idea of Progress*, pp. 67–68; William Freedman, "Swift's Struldbruggs, Progress, and the Analogy of History," *Studies in English Literature, 1500–1900* 35 (1995): 457–72. Notwithstanding the originality of Pascal's analogy, Vincent of Lerins offers a remarkable patristic parallel to do with the growth of religion in the

soul and Christian doctrine in history. *Commonitory*, ch. 23, 55–56, ANF vol. 11, p. 148. Cf. Augustine, *City of God* X.14.

10. Bacon, *English Translation of the Novum Organum, Works*, vol. 4, p. 82. For similar remarks, see Hooke, *Micrographia*, Preface, sig. d1r.; Bernard Fontenelle, *Digression on the Ancients and the Moderns*, in *Conversations with a Lady on the Plurality of Worlds*, tr. Joseph Glanvill, 4th ed. (London, 1719), pp. 177–211; Johann Gottfried Herder, *Philosophical Writings*, ed. Michael N. Forster (Cambridge: Cambridge University Press, 2002), p. 281.

11. Pascal, "Preface to Treatise on the Vacuum," p. 450. Cf. Bacon, *English Translation of the Novum Organum, Works*, vol. 4, p. 82.

12. Bacon, *English Translation of the Novum Organum*, I.81, *Works*, vol. 4, p. 80.

13. Bacon, *English Translation of the Novum Organum*, I.84, *Works*, vol. 4, p. 82; *De sapientia veterum, Works*, vol. 4, p. 753. Cf. *Parasceve, Works*, vol. 4, p. 252.

14. Sergeant, *Method to Science*, sig. d1r. The underlying assumption of the criticism is that, as Aristotle had taught, proper scientific explanation is *propter quid* (demonstrating an effect from a cause) rather than *quia* (reasoning from known effects to a cause). Aristotle, *Posterior Analytics* 1.13.

15. Locke, *Essay*, IV.xii.10 (vol. 1, p. 349). Cf. IV.iii.29 (vol. 1, p. 222), III.vi.9 (I, 64). In these contexts, Locke means "science" in the Aristotelian sense of knowledge that is certain and demonstrable. For Locke's views on the nature of natural philosophy, see Peter Anstey, "Locke on Method in Natural Philosophy," in *The Philosophy of John Locke: New Perspectives*, ed. Peter Anstey (London: Routledge, 2003), pp. 26–42.

16. On the connection between printing and progress, see A. G. Molland, "Medieval Ideas of Scientific Progress," *Journal of the History of Ideas* 39 (1978): 561–77.

17. Ann Blair, *Too Much to Know: Managing Scholarly Information before the Modern Age* (New Haven: Yale University Press, 2010); Richard Yeo. "Reading Strategies for Coping with Information Overload, ca. 1550–1700," *Journal of the History of Ideas* 64 (2003): 11–28; Daniel Rosenberg, "Early Modern Information Overload," *Journal of the History of Ideas* 64 (2003): 1–9.

18. Jonathan Swift, "Thoughts on Various Subjects, Moral and Diverting," in *Works*, vol. 14, p. 172.

19. See, e.g., Nicholas Carr, *The Shallows: How the Internet Is Changing the Way We Read, Think and Remember* (London: Atlantic Books, 2010).

20. Descartes, *Discourse on the Method*, pt. VI, CSM I, 142.

21. Jonathan Swift, *Travels into Several Remote Regions of the World* (London, 1726), pt. III, ch. 5.

22. James Harrington, *The Prerogative of Popular Government* (London, 1658), Epistle Dedicatory. Robert South: *"Mirantur nihil nisi pulices, pediculos, et seipsos,"* quoted in Isaac Disraeli, *Calamities and Quarrels of Authors* (London: Warne and Co., 1881), p. 342. On South's opposition to the Royal Society, see Larry Stewart, *The Rise of Public Science: Rhetoric, Technology, and Natural Philosophy in Newtonian Britain, 1660–1750* (Cambridge: Cambridge University Press, 1992), p. 6. See also J. R. McCulloch, ed., *Early English*

Tracts on Commerce (Cambridge: Cambridge University Press, 1952), p. 357; Samuel Butler, "Paedants," in *Satires and Miscellaneous Poetry and Prose*, ed. R. Lamar (Cambridge: Cambridge University Press, 1928), p. 166.

23. *The Diary of Samuel Pepys*, 1 February 1663/4, ed. R. Latham and W. Matthews (London: Bell, 1971), V, 33.

24. Marjorie Hope Nicholson and David Stuart Rodes, *The Virtuoso* (Lincoln: University of Nebraska Press, 1966), introduction.

25. Claude Lloyd, "Shadwell and the Virtuosi," *PMLA* 44 (1929): 472–94.

26. Thomas Shadwell, *The Virtuoso* (London, 1676), iii, 49; v, 84. For further background on literary attacks on the virtuosi, see Barbara M. Benedict, *Curiosity: A Cultural History of Early Modern Inquiry* (Chicago, 2001), pp. 46–51; Michael Hunter, *Science and Society in Restoration England* (Cambridge: Cambridge University Press, 1981), p. 111; Stephen Gaukroger, "Science, Religion and Modernity," *Critical Quarterly* 47 (2005): 1–31; Peter Harrison, "'The Fashioned Image of Poetry or the Regular Instruction of Philosophy?': Truth, Utility, and the Natural Sciences in Early Modern England," in *Science, Literature, and Rhetoric in Early Modern England*, ed. D. Burchill and J. Cummins (Aldershot: Ashgate, 2008), pp. 15–36; R. H. Syfret, "Some Early Critics of the Royal Society," *Notes and Records of the Royal Society of London* 8 (1950): 20–64.

27. Robert Boyle, *Some Considerations touching the Usefulnesse of Experimental Naturall Philosophy* (London, 1663); Thomas Sprat, *History of the Royal Society* (London, 1667); Joseph Glanvill, *Plus Ultra . . . in an Account of some of the most Remarkable Late Improvements of Practical Useful Learning* (London, 1668) and his essays "Modern Improvements of Useful Knowledge" and "The Usefulness of Real Philosophy to Religion," both in *Essays on Several Important Subjects in Philosophy and Religion* (London, 1676). See also the anonymous *Brief Vindication of the Royal Society: From the Late invectives and Misrepresentations of Mr. Henry Stubbe* (London, 1670).

28. Glanvill, "The Usefulness of Real Philosophy to Religion," p. 25.

29. Ibid., p. 5.

30. Glanvill, "Modern Improvements of Useful Knowledge," p. 35.

31. Meric Casaubon, *A Letter of Meric Casaubon . . . concerning Natural Experimental Philosophie* (Cambridge, 1669), p. 5.

32. Ibid., p. 24.

33. Ibid., p. 31.

34. Henry Stubbe, *Campanella Revived* (London, 1670), p. 14; *Plus Ultra reduced to a Non-Plus*, p. 13.

35. Meric Casaubon, *A Letter*, p. 26. See also Francis Bampfield, *All in One* (London, 1677), pp. 56–57.

36. Aquinas, ST 1a2ae, 62.

37. John Evelyn, *The Diary of John Evelyn*, 3 vols. (London: Routledge/Thoemmes, 1996), vol. 1, p. 218. Cf. *OED*, s.v. "Charity," 6, http://www.oed.com/view/Entry/30731, accessed 1 June 2012. The *OED* erroneously gives the date as 1697.

38. G. A. T. Allan, *Christ's Hospital* (Shepperton: Town and County, 1984); Albert C.

Seward, "Christ's Hospital and the Royal Society," *Notes and Records of the Royal Society of London* 3 (1940–41): 141–45.

39. See Harrison, "Curiosity, Forbidden Knowledge, and the Reformation of Natural Philosophy in Early-Modern England," *Isis* 92 (2001): 265–90.

40. Bacon, *Great Instauration, Works,* vol. 4, pp. 20 f.

41. Bacon, *Advancement of Learning, Works,* vol. 23, p. 294.

42. Bacon, *Essays, Works,* vol. 6, p. 403.

43. Bacon, *Valerius Terminus, Works,* vol. 3, pp. 221 f.

44. Joseph Glanvill, "Usefulness of Real Philosophy to Religion," p. 5, in *Essays on Several Important Subjects in Philosophy and Religion* (London, 1676), p. 25.

45. Ibid., pp. 38 f.

46. See, e.g., Robert Boyle, *Some Considerations, Works,* vol. 2, pp. 2, 5, 9, 31; Sprat, *History,* pp. 322 f.; John Edwards, *A Demonstration of the Existence and Providence of God* (London, 1696), pt. I, pp. 206–15, pt. II, p. 150.

47. Henry Stubbe, *A Censure upon certain Passages Contained in the history of the Royall Society,* 2nd ed. (Oxford, 1671), pp. 26 f.

48. Aquinas, ST 2a2ae, 23, 1.

49. On changes in conceptions of charity, see Jerome B. Schneewind, "Philosophical Ideas of Charity: Some Historical Reflections," in *Giving: Western Ideas of Philanthropy,* ed. Jerome B. Schneewind (Bloomington: Indiana University Press, 1996), pp. 54–75; Jerome B. Schneewind, "The Misfortunes of Virtue." An earlier version of this view of charity, still thoroughly theological and not wholly neglectful of the interior dimension, may be seen in John Winthrop's "A Modell of Christian Charity," where charity is described in these terms: "we must be knit together in this work as one man, we must entertain each other in brotherly affection, we must be willing to abridge ourselves of our superfluities, for the supply of others' necessities"; in *The Puritans in America: A Narrative Anthology,* ed. Andrew Delbanco and Alan Heimert (Cambridge, MA: Harvard University Press, 1985), p. 91.

50. Samuel Pufendorf, *The Whole Duty of Man According to the Law of Nature,* trans. Andrew Tooke, ed. Ian Hunter and David Saunders, with *Two Discourses and a Commentary by Jean Barbeyrac,* trans. David Saunders (Indianapolis: Liberty Fund, 2003), p. 50. In "The Misfortunes of Virtue" Schneewind suggests that the idea of "imperfect duties" provided a way of incorporating the virtues into the new schemes of natural law. That said, he points to this incorporation as a significant step in the direction of the secularization of charity.

51. Pufendorf, *The Law of Nature and of Nations,* trans. C. H. Oldfather and W. A. Oldfather (Oxford: Clarendon, 1934), 3.4.1, p. 380.

52. Pascal, *Pensées,* §201 (Lafuma).

53. Steven Weinberg, *The First Three Minutes* (New York: Basic Books, 1977), p. 144.

54. Lynn White Jr., "The Historical Roots of our Ecological Crisis," *Science* 155 (1967): 1203–7. Major discussions of the thesis include Ian Barbour, ed., *Western Man and Environmental Ethics: Attitudes towards Nature and Technology* (Reading, MA: Addison-

Wesley, 1973); Donald Gowan and Millard Shumaker, *Subduing the Earth: An Exchange of Views* (Kingston: United Church of Canada, 1980), see esp. the bibliography; Robin Attfield, *The Ethics of Environmental Concern* (Oxford: Blackwell, 1983); David and Eileen Spring, eds., *Ecology and Religion in History* (New York: Harper and Row, 1974); Carl Mitcham and Jim Grote, eds., *Theology and Technology: Essays in Christian Analysis and Exegesis* (New York: University Press of America, 1984); Jeremy Cohen, *"Be Fertile and Increase, Fill the Earth and Master It": The Ancient and Medieval Career of a Biblical Text* (Ithaca: Cornell University Press, 1989), pp. 15–18; Elspeth Whitney, "Lynn White, Eco-theology, and History," *Environmental Ethics* 15 (1993): 151–69.

55. See Harrison, "Subduing the Earth"; Cohen, *"Be Fertile and Increase."*

56. Origen, *Homiles on Leviticus*, 5.2, quoted in Patricia Cox, "Origen and the Bestial Soul," *Vigiliae Christianae* 36 (1982): 115–40 (122). See also Harrison, "Reading the Passions: The Fall, the Passions, and Dominion over Nature," in *The Soft Underbelly of Reason: The Passions in the Seventeenth Century*, ed. S. Gaukroger (London: Routledge, 1998), pp. 49–78.

57. Origen, *Homilies on Genesis and Exodus*, p. 69. For similar readings in the Middle Ages, see Cohen, *"Be Fertile and Increase,"* pp. 228 f.

58. Augustine: *Confessions* XIII.xxi (Chadwick trans. p. 291). Cf. Jerome, *Commentariorum in Hiezechielem*, 1.1.6/8 (CCSL LXXV, 11 f.).

59. Aquinas, ST 1a, 96, 2. Cf. John Chrysostum, *Homilies on Philippians*, VII (NPNF I, vol. 8); *Homilies on Titus* II (NPNF I, vol. 8); *Homilies on Timothy*, XVIII (NPNF I, vol. 8); *Select Homilies and Letters*, "To Those who had not Attended the Assembly," 4 (NPNF I, vol. 9).

60. Quoted in Merchant, *The Death of Nature* (New York: Harper and Row, 1990), p. 170. On Bacon and dominion generally, see pp. 164–90. Cf. William Leiss, *The Domination of Nature* (New York: George Braziller, 1972), pp. 45–71. On the continent, Descartes offered the similar remark that through science we can "make ourselves, as it were, the lords and masters of nature"; *Discourse on Method* VI, in CSM vol. 1, pp. 142 f.

61. *Novum Organum*, II.lii, *Works*, vol. 4, pp. 247 f. Cf. Hugh of Saint Victor's contention that the theoretical, practical, and productive sciences are to be directed to "either the restoring of our nature's integrity, or the relieving of those weaknesses to which our present life lies subject." Hugh of Saint Victor, *Didascalicon* I.v (pp. 51 f.). The *Asclepius* speaks of a twofold task of the human being: in the immortal part to imitate the divine *ratio* and *diligentia;* in the mortal part to tend and manage the earth. The latter are to be accomplished through divinely given arts and disciplines through which God intended the perfection of the world (I.viii).

62. Sprat, *History*, p. 62.

63. Joseph Glanvill, *Scepsis Scientifica* (London, 1665), Sig. b3v; *Plus Ultra* (London, 1668), p. 87 (cf. p. 104); *Scepsis Scientifica*, Sigs. b4r-v.

64. Bacon, *Valerius Terminus, Works*, vol. 3, p. 222.

65. Stephen Gaukroger, *Francis Bacon and the Transformation of Early Modern Natural Philosophy* (Cambridge: Cambridge University Press, 2001), p. 5. Cf. Antonio Pérez-Ramos, "Bacon's Legacy," *Cambridge Companion to Bacon*, pp. 311–34.

66. Seneca, *Epistles* 90, trans. Macaulay, *Critical and Miscellaneous Essays*, vol. 3, p. 437. Seneca speaks here of *sapientia* rather than *philosophia*.

67. Thomas Hobbes, *Leviathan* pt. 1, ch. xi.

68. Hobbes, *Elements of Philosophy*, in *English Works of Thomas Hobbes*, 11 vols., ed. William Molesworth, vol. 1, p. 7.

69. Robert Boyle, *Some Considerations, Works*, vol. 2, pp. 64 f., 14.

70. Aquinas, ST 1a, 62, 3.

71. Vincent of Lerins, *Commonitory*, ch. 23, 55–56, ANF vol. 11, p. 148. See also Tertullian, *De virginibus velandis* 1, PL vol. 2, p. 938; Cyril of Alexandria, *Contra Julianum IV*, PG vol. 76, pp. 694–96; Aquinas, ST 2a2ae, 1, 7. For early moderns who, exceptionally, argued for progress in religion, see Ronald S. Crane, "Anglican Apologetics and the Idea of Progress," *Modern Philology* 31 (1934): 273–306. The best-known modern treatment of this topic is John Henry Newman, *Essay on the Development of Christian Doctrine* (London: James Toovey, 1845).

72. Cotton Mather, *American Tears upon the Ruines of the Greek Churches* (Boston, 1701), pp. 42 f.; Francis Bacon, *The Advancement of Learning*, ed. Arthur Johnston (Oxford: Clarendon Press, 1974), 42; Sprat, *History*, p. 371. See also Thomas Culpeper, *Morall Discourses and Essayes* (London, 1655), 63; Samuel Hartlib Sheffield University Library, Hartlib Papers XLVIII 17, reproduced in Webster *The Great Instauration*, Appendix 1, 524–28; Noah Biggs, *Mataeotechnia Medicinae Praxeos. The Vanity of the Craft of Physick* (London, 1651), To the Parliament.

73. Nicolas de Condorcet, *Sketch for a Historical Picture of the Progress of the Human Mind*, trans. June Barraclough (New York: Noonday, 1955), p. 72.

74. For these themes, see Webster, *The Great Instauration*; Harrison, *The Bible and the Rise of Science*, ch. 6; *Fall of Man*, ch. 5.

75. Larry Laudan, *Progress and Its Problems* (London: Routledge, 1977), ch. 4. There is, admittedly, the possibility that some utopian end, though practically unachievable, might serve as a regulative ideal. See Ilkka Niiniluoto, *Is Science Progressive?* (Dordrecht: Kluwer, 1984), esp. ch. 5.

Chapter Six

1. W. G. Ward, "Science, Prayer, Free Will, and Miracles," *Dublin Review*, 8/16 April 1867, 255–98 (255 n.).

2. See "science," n., 5b, *OED*, http://www.oed.com/view/Entry/172672, accessed 18 May 2012.

3. The story related here focuses very much on the English context. For a parallel account of the emergence of *Naturwissenschaft* (natural science) as a new category in nineteenth-century Germany, see Denise Phillips, *Acolytes of Nature: Defining Natural Science in Germany, 1770–1850* (Chicago: University of Chicago Press, 2012).

4. Edward B. Freeland, "Buckle, Draper, and a Science of History," *Continental Monthly* 4 (July–September, 1863): 610–24 (610).

5. William Whewell, *The Philosophy of the Inductive Sciences: Founded upon Their History*, new ed., 2 vols. (London: John Parker, 1847), vol. 2, p. 117; *History of the Induc-*

tive Sciences, vol. 2, p. 571; William Kirby, *On the History, Habits and Instincts of Animals* (Philadelphia, 1837), p. xxi [(London, 1835), p. i]; Review of T. Chalmers, *The Evidence and Authority of Christian Revelation, Christian Observer* 14 (1815): 247–48 (249). Cf. Brooke Foss Westcott, *The Gospel of Life* (London: Macmillan, 1892), p. 89.

6. *Encyclopaedia Britannica* (Edinburgh, 1771), 3 vols., s.v. "Science" (III, 570a). Cf. Chambers, *Cyclopaedia; or, An universal dictionary of arts and sciences* (London, 1728), 2 vols., vol. 1, p. vii; William Thomson, *Outline of the Laws of Thought* (London: Pickering, 1842), pp. 26–38.

7. E. A. Freeman, *Letters* (MS.), 10 February, 1884, quoted in *OED*, s.v. "Science," 5c, http://www.oed.com/view/Entry/172672, accessed 29 May 2012.

8. Mark Pattison, *Essays by the Late Mark Pattison* (Oxford: Clarendon, 1889), p. 468.

9. "The Late Warden of New College," *Athenæum*, 7 February 1903, No. 3928, p. 176.

10. Adrian Desmond, *Huxley*, 2 vols. (London: Michael Joseph, 1997), II, xiii, 3.

11. See, e.g., Andrew Cunningham, "How the Principia Got Its Name; or, Taking Natural Philosophy Seriously," *History of Science* 28 (1991): 377–92; Edward Grant, "God and Natural Philosophy: The Late Middle Ages and Sir Isaac Newton," *Early Science and Medicine* 6 (2000): 279–98; Peter Dear, "Religion, Science, and Natural Philosophy: Thoughts on Cunningham's Thesis," *Studies in History and Philosophy of Science* 32 (2001): 377–86; Harrison, "Physico-Theology and the Mixed Sciences."

12. Noël-Antoine Pluche, *Spectacle de la Nature; or, Nature Display'd*, 7 vols. (London, 1770) vol. 3, p. 112, cf. p. 304.

13. Martha McMacklin Garland, *Cambridge before Darwin: The Ideal of Liberal Education, 1800–1860* (Cambridge: Cambridge University Press, 1980), pp. 52–58. Cf. Julie A. Reuben, *The Making of the Modern University* (Chicago: University of Chicago Press, 1996), pp. 22 f.

14. John Yolton, "Schoolmen, Logic and Philosophy," in *The History of the University of Oxford*, vol. 5, *The Eighteenth Century*, ed. L. S. Sutherland and L. G. Mitchell (Oxford: Clarendon, 1986), pp. 565–91.

15. Garland, *Cambridge before Darwin*, pp. 90, 108 f.

16. Paley, *Natural Theology, Paley's Works* (London: Bohn, 1849), p. 113.

17. *The Bridgewater Treatises on the Power, Wisdom, and Goodness of God, as Manifested in the Creation.* Treatise V, Peter Roget, *Animal and Vegetable Physiology Considered with Reference to Natural Theology*, 2 vols. (Philadelphia: Carey, Lea & Blanchard, 1836), vol. 1, pp. viii, ix.

18. Quoted in Jonathan Topham, "Science and Popular Education in the 1830s: The Role of the *Bridgewater Treatises*," *British Journal for the History of Science* 25 (1992): 397–430 (414).

19. Francis Wayland, *The Elements of Moral Science* (New York: Cooke and Co., 1835), p. 127.

20. Christopher Brooke, *A History of the University of Cambridge IV, 1870–1990* (Cambridge: Cambridge University Press, 1988), p. 158.

21. John Henslow, *Questions on the Subject Matter of Sixteen Lectures in Botany*

Required for a Pass Examination (Cambridge: Cambridge University Press, 1851). See also David Layton, *Science for the People* (London: George Allen and Unwin, 1973).

22. Reuben, *Making of the Modern University,* ch. 5.

23. Brooke and Cantor, *Reconstructing Nature,* pp. 156 f.

24. Tony Bennett, *The Birth of the Museum: History, Theory, Politics* (London: Routledge, 1995), p. 28.

25. Glanvill, "Usefulness of Real Philosophy," p. 5; Sprat, *History of the Royal Society,* pp. 349 f., 366 f. See also Harrison, "The Cultural Authority of Natural History in Early Modern Europe," in *Biology and Ideology from Descartes to Dawkins,* ed. Denis Alexander and Ronald Numbers (Chicago: University of Chicago Press, 2010), pp. 11–35. Marin Mersenne, *La vérité des sciences* (Paris, 1625), Preface.

26. John Herschel, *Preliminary Discourse,* new ed. (London: Longman, Brown, Green & Longmans, 1851), pp. 5, 16.

27. George Fairholme, *A General View of the Geology of Scripture* (London: James Ridgway, 1833), p. 28. Cf. Henry Brougham, who characterized natural theology as "an exercise at once intellectual and moral, in which the highest faculties of the understanding and the warmest feelings of the heart alike partake." *Natural Theology* (London, 1835), p. 125. See Robert N. Proctor, *Value-Free Science? Purity and Power in Modern Knowledge* (Cambridge, MA: Harvard University Press, 1991); Richard Yeo, *Defining Science: William Whewell, Natural Knowledge and Public Debate in Early Victorian Britain* (Cambridge: Cambridge University Press, 1993).

28. Herschel, *Preliminary Discourse,* p. 14.

29. William Kirby, *On the History, Habits and Instincts of Animals* (Philadelphia: Carey, Lea and Blanchard, 1837), p. xxi [(London, 1835), p. i]; Whewell, *History of the Inductive Sciences,* vol. 2, p. 571. Cf. J. W. Clark and T. Hughes, eds., *The Life and Letters of the Rev. Adam Sedgwick* (Cambridge: Deighton, Bell and Co., 1890), vol. 1, p. 362.

30. Robert M. Young, "Natural Theology, Victorian Periodicals, and the Fragmentation of a Common Context," in *Darwin's Metaphor: Nature's Place in Victorian Culture* (Cambridge: Cambridge University Press, 1985), pp. 126–63. For discussions of the thesis, see John Hedley Brooke, "Natural Theology and the Plurality of Worlds: Observations on the Brewster-Whewell Debate," *Annals of Science* 34 (1977): 221–86; Richard Yeo, "The Principle of Plenitude and Natural Theology in Nineteenth-Century Britain," *British Journal for the History of Science* 19 (1986): 263–82; Jonathan Topham, "Beyond the 'Common Context': The Production and Reading of the Bridgewater Treatises," *Isis* 89 (1998): 233–62.

31. See Whewell's classification of the sciences in *Philosophy of the Inductive Sciences,* vol. 2, p. 117.

32. Anon., "Crombie's *Natural Theology,*" *Quarterly Review* 51 (March and June 1834): 216–17.

33. Ibid., pp. 216, 218.

34. Leonard Woods, *Letters to Unitarians and Reply to Dr. Ware* (Andover: Mark Newton, 1822), pp. 18 f.

35. J. S. Lamar, *The Organon of Scripture; or, The Inductive Method of Biblical Inter-pretation* (Philadelphia: J. B. Lippincott, 1860), pp. iii, vii, ch. 2 passim. See also David Roberts Dungan, *Hermeneutics: A Text Book* (Cincinnati: Standard Publishing Co., 1888); Clinton Lockhart, *Principles of Interpretation as Recognized Generally by Biblical Scholars, Treated as a Science, Derived Inductively from an Exegesis of Many Passages of Scripture*, 2nd ed., rev. (Fort Worth: S. H. Taylor, 1915).

36. J. Angus, *Theology an Inductive and Progressive Science* (London, n.d.), pp. 20 f., cited in D. W. Bebbington, *Evangelicalism in Modern Britain: A History from the 1730s to the 1980s* (London: Unwin Hyman, 1989), pp. 143 f.

37. Theodore Bozeman, *Protestants in an Age of Science: The Baconian Ideal and Ante-bellum American Religious Thought* (Chapel Hill: University of North Carolina Press, 1977), p. 72. See also George Marsden, *Understanding Fundamentalism and Evangelical-ism* (Grand Rapids: Eerdmans, 1991), pp. 122–52; Mark Noll, "Evangelicalism and Fun-damentalism," in *Science and Religion*, ed. Gary Ferngren (Baltimore: Johns Hopkins, 2002), pp. 261–76.

38. Both cited in George Marsden, "Understanding Fundamentalist Views of Sci-ence," in *Science and Creationism*, ed. A. Montagu (Oxford: Oxford University Press, 1984), p. 98.

39. Charles Hodge, *Systematic Theology*, 3 vols., [1872–73] I, 1, §5, §6 (Grand Rapids: Eerdmans, 1970), vol. 1, pp. 9–17.

40. Paley, *Evidences, Works*, p. 157. By "moral evidence," Paley presumably means physical evidence that leads to "moral" rather than demonstrative certainty.

41. Paley, *Natural Theology, Works*, pp. 19 f.

42. Review of Olinthus Gregory, *Letters to a Friend, on the Evidences, Doctrines, and Duties of the Christian Religion*, 2 vols. (London, 1812), in *Christian Observer* 11 (1812): 577–87 (577).

43. William Whewell, *Indications of the Creator*, 2nd ed. (London: John Parker, 1846), p. 34.

44. Richard Yeo, "William Whewell, Natural Theology and the Philosophy of Sci-ence," *Annals of Science* 26 (1979): 493–512.

45. Yeo, "Scientific Method and the Image of Science," p. 67.

46. Roget, *Animal and Vegetable Physiology*, vol. 1, pp. 20 f.

47. Kirby, *History, Habits and Instincts of Animals*, pp. xxxvii, xxii.

48. Whewell, *History of the Inductive Sciences*, vol. 2, p. 106. For further discussion, see Yeo, *Defining Science*, pp. 120–24.

49. Edmund Burke, *Reflections on the Revolution in France* [1790] (New York, 1909–14). See also Ursula Henriques, *Religious Toleration in England, 1787–1833* (Toronto: Uni-versity of Toronto Press, 1961); V. Kiernan, "Evangelicalism and the French Revolution," *Past and Present* 1 (1952): 44–56.

50. Brooke and Cantor, *Reconstructing Nature*, pp. 156 f.

51. Whewell, *On the Principles of the English University Education* (London, 1837), pp. 48 f.

52. [Adam Sedgwick], Review of "Vestiges of the Natural History of Creation," *Edinburgh Review* 82 (July 1845): 85, cf. p. 2.

53. See, e.g., Adam Sedgwick, "Objections to Mr Darwin's *Origin of Species*," *The Spectator*, March 24, 1860; T. R. Birks, *The Scripture Doctrine of Creation* (London: SPCK, 1872). Cf. G. Frederick White, Review of *Darwinism and Other Essays*, by John Fiske, *Bibliotheca Sacra* 36 (1879): 784, quoted in Ronald Numbers, *The Creationists* (Berkeley, 1993), p. 27.

54. Glanvill, *Philosophia Pia*, p. 46. See also Corneanu, *Regimens of the Mind*, pp. 79–106; Lorraine Daston, "Attention and the Values of Nature in the Enlightenment," in *The Moral Authority of Nature*, ed. Lorraine Daston and Fernando Vidal (Chicago: University of Chicago Press, 2004), pp. 100–126.

55. Jonathan Edwards, *A Treatise concerning Religious Affections* [1746], in *The Works of Jonathan Edwards*, 2 vols. (Peabody: Hendrickson, 1998), vol. 1, p. 333. See also Harrison, "Experimental Religion and Experimental Science in Early Modern England," *Intellectual History Review* 21 (2011): 413–33.

56. Paley, *Natural Theology, Works*, p. 135. It has been argued that Bacon himself saw induction as a kind of meditative process. See Michael Hattaway, "Bacon and 'Knowledge Broken': Limits of Scientific Method," *Journal of the History of Ideas* 39 (1978): 183–97 (188).

57. T. Chalmers, "On Mechanic Schools, and on Political Economy as a Branch of Popular Education," in *On the Christian and Civic Economy of Large Towns*, 3 vols. (Glasgow, 1821–26), vol. 3, pp. 378–408 (378 f.), quoted in J. Topham, "Science and Education," p. 406.

58. Max Weber, "Science as a Vocation," in *From Max Weber: Essays in Sociology*, ed. H. H. Gert and C. Wright Mill (London: Routledge, 1970), pp. 142–43.

59. [William Whewell], "Mrs Sommerville on the Connexion of the Physical Sciences," *Quarterly Review* 51 (March and June 1834): 54–68; 59–60, 58–59.

60. *The Works of Samuel Taylor Coleridge, Prose and Verse* (Philadelphia, 1846), p. 512.

61. [James Burton Robertson], "Science and Revealed Religion," *Dublin Review*, April 1837, 293–329 (295 f.).

62. Whewell, "Connexion of the Physical Sciences," p. 59. Cf. Charles Babbage, *The Exposition of 1851* (London: John Murray, 1851), p. 189. For much of this paragraph I have drawn upon Sydney Ross, "'Scientist': The Story of a Word," *Annals of Science* 18 (1962): 65–86.

63. Whewell, "Connexion of the Physical Sciences," pp. 59 f. Ross, "Scientist," p. 76.

64. "Origin and Objects of the British Association," *Proceedings of the Fifth Meeting of the British Association for the Advancement of Science*, 2nd ed. (Dublin: Philip Dixon Hardy, 1835), pp. 5–10.

65. Patrick Armstrong, *The English Parson-Naturalist: A Companionship between Science and Religion* (Leominster, Herefordshire: Gracewing, 2000), p. 145.

66. Adrian Desmond and James R. Moore, *Darwin* (London: Michael Joseph, 1991), p. 526.

67. J. Vernon Jensen, *Thomas Henry Huxley, Communicating for Science* (London: Associated University Press, 1991), p. 143.

68. J. Vernon Jenson, "The X-Club: Fraternity of Victorian Scientists," *British Journal for the History of Science* 5 (1970): 63–72; Ruth Barton, "'An Influential Set of Chaps': The X-Club and Royal Society Politics, 1864–85," *British Journal for the History of Science* 23 (1990): 53–81.

69. Barton, "'An Influential Set of Chaps'"; T. W. Heyck, *The Transformation of Intellectual Life in Victorian England* (London: Croom Helm, 1982), p. 87.

70. Francis Galton, *English Men of Science: Their Nature and Nurture* (London: Macmillan, 1874), pp. 24, 26.

71. Thomas Beddoes, *Contributions to Physical and Medical Knowledge* (Bristol, 1799), p. 4.

72. For an earlier Latin reference, see vol. 3 of Michael Christoph Hanow, *Philosophiae naturalis sive physicae dogmaticae* (1766), which is entitled *"Geologia, biologia, phytologia generalis et dendrologia."* Cf. William Lawrence, *Lectures on physiology, zoology, and the natural history of man* [1819] (Salem, 1828), p. 59, who cites Treviranus. Beddoes, however, clings to the older idea that biology "is the foundation of ethics and pneumatology." See also Pietro Corsi, "Biologie," in *Lamarck, philosophe de la nature*, ed. Pietro Corsi, Jean Gayon, Gabriel Gohau, and Stéphane Tirard (Paris: Presses Universitaires de France, 2006), pp. 37–64.

73. Whewell, *Philosophy of the Inductive Sciences*, vol. 1, p. 544.

74. Huxley to Hooker, 29 July 1859, in Leonard Huxley, *The Life and Letters of Thomas Henry Huxley*, 3 vols. (New York: D. Appleton, 1900), vol. 1, p. 177.

75. Harrison, "Natural History," in Harrison, Numbers, and Shank, eds., *Wrestling with Nature*, pp. 117–48.

76. W. Stanley Jevons, *The Principles of Science: A Treatise on Logic and the Scientific Method* (London: Macmillan, 1874). On the idea of the scientific method in nineteenth-century United States, see Daniel P. Thurs, "Scientific Methods" in Harrison, Numbers, and Shank, eds., *Wrestling with Nature*, pp. 307–35.

77. R. Adamson, "Bacon," *Encyclopaedia Britannica*, 9th ed., 25 vols. (Edinburgh, 1875–89), vol. 3, pp. 200–18 (p. 216). Quoted in Yeo, "An Idol of the Market-Place," 280.

78. For an account of these changes in the sphere of experimental physics, see, e.g., K. Caneva, "From Galvanism to Electrodynamics: The Transformation of German Physics and Its Social Context," *Historical Studies in the Physical Sciences*, 9 (1978): 63–159.

79. "Professor Huxley on Men of Science," *The Mechanics' Magazine*, 14 October 1871, 284–85 (284). Huxley was reporting on a speech by the geologist T. H. Cope.

80. Robert K. Merton, *The Sociology of Science: Theoretical and Empirical Investigations* (Chicago: University of Chicago Press, 1978), pp. 277–78. For a qualified rejection of this notion of moral equivalence, see Steven Shapin, *The Scientific Life*. This is not the place for sustained engagement with Shapin's intriguing thesis, but three brief points can be made. First, attitudes such as trust are an important element of any successful organization, and not exclusively scientific ones. Second, following on from this, any

argument for specific scientific virtues assumes that the profession "scientist" and the activities that can be labeled "science" are genuinely distinctive. Third, even if these scientific virtues did exist, they function as prerequisites for scientific activity and for its successful outcomes. In the premodern and early modern periods, natural philosophy was pursued at least in part to produce the requisite virtues in its practitioners rather than the other way around.

81. Jon Roberts, "Science and Religion," in Harrison, Numbers, and Shank, eds., *Wrestling with Nature*, pp. 253–79.

82. Nicholas Wiseman, *Lectures on the Connexion between Science and Revealed Religion*, 2 vols. (London: Booker, 1836).

83. Draper, *History of the Conflict*, pp. 52, 157–59, 160 f., 168 f.; White, *History of the Warfare*, vol. 1, pp. 71–74, 108, 118; vol. 2, pp. 49–55; 55–63. For a debunking of many of these myths, see Numbers, *Galileo Goes to Jail*. Specifically on the flat earth, see Jeffrey Burton Russell, *Inventing the Flat Earth: Columbus and Modern Historians* (New York: Praeger, 1991).

84. Draper, *History of the Conflict*, p. vi; White, *History of the Warfare*, vol. 1, p. ix.

85. Ibid., pp. vi, 286.

86. Ibid., pp. 246 f.

87. *Auguste Comte and Positivism*, ed. Lenzer, pp. 71–86. Those with religious convictions also found the notion of progress favorable. See Whewell, *History of the Inductive Sciences*, vol. 1, pp. 4 f., 9 f.

88. Draper, *History of the Conflict*, pp. 364, 367.

89. A. W. Benn, *A History of English Rationalism in the Nineteenth Century* (London: Longmans, Green, and Todd, 1906), vol. 1, p. 198.

90. Barbour, *Religion and Science*, pp. 77–105. For some difficulties with the typology, see Geoffrey Cantor and Chris Kenny, "Barbour's Fourfold Way: Problems with His Taxonomy of Science-Religion Relationships," *Zygon* 36 (2001): 765–81; Mikael Stenmark, "Ways of Relating Science and Religion," in Harrison, *Cambridge Companion to Science and Religion*, pp. 278–194. See also Robert J. Russell, "Ways of Relating Science and Religion," http://www.counterbalance.org/rjr/atypo-body.html, accessed 15 July 2012.

91. The phrase "by the Creator" was added in the second and subsequent editions. Online Variorum of Darwin's *Origin of Species*: first British edition (1859), p. 490, http://darwin-online.org.uk/Variorum/1859/1859-490-c-1860.html#, accessed 18 June 2012.

92. E. O. Wilson, *On Human Nature* (Cambridge, MA: Harvard University Press, 1978), pp. 192, 201.

93. Eric Chaisson, *Epic of Evolution: Seven Ages of the Cosmos* (New York: Columbia University Press, 2006); David Christian, *Maps of Time: An Introduction to Big History* (Berkeley: University of California Press, 2004), pp. 1–14.

94. E. O. Wilson, *The Creation: An Appeal to Save Life on Earth* (New York: Norton, 2006), pp. 104 f., 165.

95. Loyal G. Rue, *Everybody's Story: Wising Up to the Epic of Evolution* (Albany: SUNY Press, 2000), p. xiii.

96. Sam Harris, *The Moral Landscape* (London: Transworld, 2010), pp. 2–3, 6–7; "Richard Dawkins Wants Evolutionary Science to Become the New Classics," http:// www.guardian.co.uk/science/blog/2012/jun/12/richard-dawkins-evolution-new -classics?INTCMP=SRCH, accessed 25 June 2012. Cf. Richard Dawkins, "Is Science a Religion?," *The Humanist*, January–February 1977, http://www.thehumanist.org /humanist/articles/dawkins.html, accessed 25 June 2012.

97. Christopher Hitchens, *God Is Not Great*, p. 94.

98. Richard Dawkins, *The God Delusion*, p. 307. In Christian theology, eschatology and ethics are often closely linked. See, e.g., Jürgen Moltmann, *The Theology of Hope* (Minneapolis: Fortress Press, 2012), p. 9.

99. Steven Pinker, *The Better Angels of Our Nature: Why Violence Has Declined* (New York: Viking, 2011).

100. For an account of these dual aspects of science, and their relations, see Peter Dear, *The Intelligibility of Nature: How Science Makes Sense of the World* (Chicago: University of Chicago Press, 2006), esp. pp. 1–14.

Epilogue

1. Niall Ferguson, *Civilization: The West and the Rest* (London: Penguin Books, 2012), p. 18.

2. Nicolai Berdyaev, *Slavery and Freedom* (London: Centenary Press, 1947), p. 9.

3. For representative discussions, see Quentin Skinner, "Meaning and Understanding in the History of Ideas," *History and Theory* 8 (1969): 3–53; Mark Bevir, *The Logic of the History of Ideas* (Cambridge: Cambridge University Press, 1999); Iain Hampsher-Monk, Karin Tilmans, and Frank van Vree, eds., *History of Concepts: Comparative Perspectives* (Amsterdam: Amsterdam University Press, 1998); David Armitage, "What's the Big Idea? Intellectual History and the *Longue Durée*," *History of European Ideas* 38 (2012): 493–507.

4. See Mark Bevir, "The Errors of Linguistic Contextualism," *History and Theory* 31 (1992): 276–98.

5. John Donne, "An Anatomy of the World."

6. Rob Iliffe, "Abstract Considerations: Disciplines and the Incoherence of Newton's Natural Philosophy," *Studies in History and Philosophy of Science* 35A (2004): 427–54. Cf. Peter Dear, "Method and the Study of Nature," *Cambridge History of Seventeenth-Century Philosophy* vol. 1, pp. 147–77 (166).

7. See, e.g., John Dupré, *The Disorder of Things: Metaphysical Foundations of the Disunity of Science* (Cambridge, MA: Harvard University Press, 1993). Peter Galison and David J. Stump, eds., *The Disunity of Science: Boundaries, Contexts and Power* (Stanford: Stanford University Press, 1996); Nancy Cartwright, *The Dappled World: A Study of the Boundaries of Science* (Cambridge: Cambridge University Press, 1999).

8. Locke, *Essay*, II.xxiii.12 (I, 402). For more on this theme, see Harrison, *Fall of Man*, pp. 245–58.

9. In stressing the significance of the Protestant Reformation and the role of the Fall in the shaping of Western modernity, my emphasis differs from those who have

sought to locate the origins of modernity in the emergence of medieval nominalism and voluntarism. See ch. 3, n. 94.

10. Gaukroger, *Emergence of a Scientific Culture*, p. 23.

11. Ibid., p. 11.

12. Taylor, *A Secular Age*, pp. 20, p. 159 f.

13. Charles Taylor, "The Politics of Recognition," in *Multiculturalism: Examining the Politics of Recognition*, ed. Amy Gutman (Princeton: Princeton University Press, 1994), pp. 25–73; Alasdair MacIntyre, *Whose Justice? Which Rationality?* (London: Duckworth, 1988).

14. Max Weber, *The Protestant Ethic and the Spirit of Capitalism* [1904–5], trans. Talcott Parsons (London: Routledge, 2001), p. xxviii.

15. Gaukroger, *The Emergence of a Scientific Culture*, p. 11.

16. Gotthold Lessing, "On the Proof of the Spirit and of Power," in *Lessing's Theological Writings*, ed. Henry Chadwick (Stanford: Stanford University Press, 1957), p. 53. A similar principle informed seventeenth-century doubts about whether Baconian natural histories could provide the basis for a proper science. See, e.g., John Sergeant, *The Method to Science* (London, 1696), Preface. Cf. Aristotle, *Metaphysics* 1027a19–21.

17. For a perceptive linking of the Tübingen school of Baur with the history of science, see David Bloor, "Rationalism, Supernaturalism, and the Sociology of Knowledge," in *Scientific Knowledge Socialized: Selected Proceedings of the 5th Joint International Conference on the History and Philosophy of Science Organized by the IUHPS, Veszprém, 1984*, ed. I. Hronszk, M. Fehér, and B. Dajka (Dordrecht: Kluwer, 1988), pp. 59–74, and Bloor, "Epistemic Grace: Antirelativism as Theology in Disguise," *Common Knowledge* 12 (2007): 250–80.

18. Arguably Lessing himself had a similar understanding of a kind of universal religion. See Toshimasa Yasukata, *Lessing's Philosophy of Religion and the German Enlightenment* (New York: Oxford University Press, 2002), pp. 62–82.

19. Larry Laudan, *Progress and Its Problems: Towards a Theory of Scientific Growth* (London: Routledge and Kegan Paul, 1977), p. 189. Cf. Gaukroger: "Science is thereby protected in advance from the historicization and contextualization that, coming to a head in the middle of the nineteenth century, eventually undermined Christianity's claims to a *sui generis* legitimacy." *Emergence of a Scientific Culture*, pp. 11–12.

20. Bloor, "Rationalism, Supernaturalism, and the Sociology of Knowledge"; James A. Moore, "Cutting Both Ways—Darwin among the Devout: A Response to David Livingstone, Sara Miles, and Mark Noll," *Perspectives on Science and Christian Faith* 45 (1994): 169.

21. Philosopher of science Hans Reichenbach thus claimed that "science has taken over a social function which originally was satisfied by religion: the function of offering ultimate security. The belief in science has replaced, in large measure, the belief in God." *The Rise of Scientific Philosophy* (Berkeley: University of California Press, 1951), pp. 43 f. Cf. Michael Gillespie, for whom modernity involves not so much the erasure of God "as the transfer of his attributes, essential powers and capacities to other entities or realms of being." *Theological Origins of Modernity*, p. 274.

22. Christopher Hitchens, *God Is Not Great: The Case against Religion* (London, 2007). The list comes from Preston Jones's review of Christopher Hitchens, *God Is Not Great*, in *Books and Culture* (2007).

23. Sam Harris, "Science Must Destroy Religion," in *What Is Your Dangerous Idea?*, ed. John Brockman (New York: Harper, 2007), pp. 148–51.

24. David Hume, *The Natural History of Religion*, ed. H. E. Root (Stanford: Stanford University Press, 1957), p. 21.

25. Pascal Boyer, *Religion Explained: The Human Instincts That Fashion God, Spirits and Ancestors* (London: Heinemann, 2001); Daniel Dennett, *Breaking the Spell: Religion as a Natural Phenomenon* (London: Penguin, 2007).

26. As for theology, as opposed to religion, the modern West has inherited a conception of faith that is to some degree distorted by the epistemological paradigm that has also afflicted modern philosophy. Viewed from a historical perspective, the two traditional components of faith—*fiducia* (trust) and *assensus* (intellectual assent)—have been reduced to the latter alone. This eclipse of *fiducia* is surely a vital element of what Charles Taylor has referred to as "new conditions for belief" in the modern age. *A Secular Age*, pp. 20, p. 159 f.

27. On this theme, see William T. Cavanaugh, *The Myth of Religious Violence* (Oxford: Oxford University Press, 2009). See also José Casanova, "The Secular, Secularizations, Secularisms," in Calhoun, Juergensmeyer, and VanAntwerpen, eds., *Rethinking Secularism*, pp. 54–74. It follows, as already implied in ch. 4, that it was not religion and the conflicts that it caused that necessitated the Westphalia settlement, but rather that the Westphalia Treaty was the occasion for the emergence of the modern conception "religion," making possible the anachronistic notion of "wars of religion."

28. Samuel P. Huntington, *The Clash of Civilizations and the Remaking of World Order* (New York: Simon and Schuster, 1998). Cf. Tzvetan Todorov, *The Fear of Barbarians: Beyond the Clash of Civilizations* (Chicago: University of Chicago Press, 2010).

[BIBLIOGRAPHY]

Primary Sources

Abercrombie, D. *Academia Scientiarum; or, The Academy of the Sciences, Being a Short and Easy Introduction to the Knowledge of Liberal Arts and Sciences.* London, 1687.

Adamson, R. "Bacon." In *Encyclopaedia Britannica*, vol. 3, pp. 200–18. 9th ed. 25 vols. Edinburgh, 1875–89.

Addison, Joseph. *The Evidences of the Christian Religion* [1721]. London, 1807.

Agrippa, Heinrich Cornelius. *Three Books of Occult Philosophy.* London, 1641.

Albertus Magnus. *Alberti Magni Opera Omnia edenda curavit Institutum Alberti Magni Coloniense Bernhardo Geyer praeside.* 38 vols. Münster: Aschendorff, 1951–.

Ames, William. *The Marrow of Sacred Divinity.* London, 1642.

Anaxagoras. *Anaxagoras of Clazomenae: Fragments and Testimonia: A Text and Translation.* Translated by Patricia Curd. Toronto: University of Toronto Press, 2007.

Anon. *Brief Vindication of the Royal Society: From the Late invectives and Misrepresentations of Mr. Henry Stubbe.* London, 1670.

———. "Crombie's *Natural Theology.*" *Quarterly Review* 51 (March and June 1834): 216–17.

———. "The Late Warden of New College." *Athenæum* 3928 (7 February 1903): 176.

———. "Origin and Objects of the British Association." *Proceedings of the Fifth Meeting of the British Association for the Advancement of Science.* 2nd ed. Dublin: Philip Dixon Hardy, 1835.

———. "Professor Huxley on Men of Science." *The Mechanics' Magazine.* 14 October 1871, 284–85.

———. Review of *The Evidence and Authority of Christian Revelation*, by T. Chalmers. *Christian Observer* 14 (1815): 247–48.

———. Review of *Letters to a Friend, on the Evidences, Doctrines, and Duties of the Christian Religion*, by Olinthus Gregory. *Christian Observer* 11 (1812): 577–87.

Aristotle. *Complete Works of Aristotle*. 2 vols. Edited by Jonathan Barnes. Princeton: Princeton University Press, 1984.

Arnim, Hans von, ed. *Stoicorum Veterum Fragmenta*. 4 vols. Stutgardiae: B. G. Teubneri, 1964.

Augustine. *Augustine: Earlier Writings*. Edited by John H. S. Burleigh. London: SCM, 1953.

———. *City of God*. 2 vols. Translated by Marcus Dodds. New York: Random House, 1950.

———. *The Complete Works of St. Augustine*. Edited by Boniface Ramsey. New York: New City Press, 1991–.

———. *The Profit of Believing: Very Usefull Both for All Those That Are Not Yet Resolved What Religion They Ought to Embrace: And For Them That Desire to Know Whither Their Religion Be True or No*. Translated by A. P. London, 1651.

———. *Saint Augustine's Confessions*. Translated by Henry Chadwick. Oxford: Oxford University Press, 1991.

Avicenna (Ibn Sīnā). *Avicenna Latinus. Liber de philosophia prima sive scientia divina*. 8 vols. Edited by Simone van Riet. Louvain-Leiden: Brill, 1968–72.

Babbage, Charles. *The Exposition of 1851*. London: John Murray, 1851.

Bacon, Francis. *The Advancement of Learning*. Edited by Arthur Johnston. Oxford: Clarendon Press, 1974.

———. *The Works of Francis Bacon*. 14 vols. Edited by James Spedding, Robert Ellis, and Douglas Heath. London: Longman, 1857–74.

Bacon, Roger. *Opus Majus*. 2 vols. Translated by Robert Belle Burke. Philadelphia: University of Pennsylvania Press, 1928.

Bampfield, Francis. *All in One*. London, 1677.

Barrow, Isaac. *The Works of Isaac Barrow*. 3 vols. New York: John Riker, 1845.

Barth, Karl. *Church Dogmatics*. 4 vols. Edinburgh: T. & T. Clark, 1936–69.

Baxter, Richard. *Reasons of the Christian Religion*. London, 1667.

Beddoes, Thomas. *Contributions to Physical and Medical Knowledge*. Bristol, 1799.

Benn, A. W. *A History of English Rationalism in the Nineteenth Century*. 2 vols. London: Longmans, Green, and Todd, 1906.

Bentley, Richard. *The Works of Richard Bentley, D.D.* Edited by Alexander Dyce. Vol. 3. London: Macpherson, 1838.

Bernard, Richard. *Look Beyond Luther; or, An Answere to that Question, So Often and So Insultingly Proposed by Our Adversaries, Asking Us; where this Our Religion was Before Luther's Time*. London, 1623.

Biggs, Noah. *Mataeotechnia Medicinae Praxeos. The Vanity of the Craft of Physick*. London, 1651.

Birks, T. R. *The Scripture Doctrine of Creation*. London: SPCK, 1872.

Bodin, Jean. *Colloquium of the Seven about the Secrets of the Sublime*. Translated by Marion Kurtz. Princeton: Princeton University Press, 1975.

Bonaventure. *Commentaria in Quatuor Libros Sententiarum, Opera Omnia S. Bonaventurae.* 10 vols. Quaracchi: Ad Claras Aquas, 1882–1902.

Bonhoeffer, Dietrich. *Letters and Papers from Prison.* New York: Macmillan, 1962.

Boyle, Robert. Boyle Papers, Library of the Royal Society, London.

———. *Some Considerations Touching the Usefulnesse of Experimental Naturall Philosophy.* London, 1663.

———. *Some Physico-Theological Considerations about the Possibility of the Resurrection.* London, 1675.

———. *The Works of Robert Boyle.* 14 vols. Edited by Michael Hunter and Edward Davis. London: Pickering & Chatto, 1999–2000.

———. *Works of the Honourable Robert Boyle.* 6 vols. Edited by Thomas Birch. Hildesheim: Georg Olms, 1966.

Brougham, Henry. *A Discourse of Natural Theology: Showing the Nature of the Evidence and the Advantages of the Study.* London, 1835.

Bullinger, Heinrich. *Fiftie Godly and Learned Sermons.* London, 1577.

Bunsen, Christian. *Hipploytus and His Age.* 4 vols. London: Longman, Brown, Green and Longmans, 1852.

Burgersdijck, Franco. *Monitio Logica.* London, 1697.

Burke, Edmund. *Reflections on the Revolution in France* [1790]. New York: P. F. Collier & Son, 1909–14.

Butler, Samuel. *Characters and Passages from Notebooks.* Edited by A. R. Waller. Cambridge: Cambridge University Press, 1908.

———. "Paedants." In *Satires and Miscellaneous Poetry and Prose,* edited by R. Lamar, 166. Cambridge: Cambridge University Press, 1928.

Byfield, Nicholas. *The Rule of Faith; or, An Exposition of the Apostles Creed.* London, 1626.

Calvin, John. *Calvin's Commentaries.* 22 vols. Grand Rapids: Baker Books, 2003.

———. *Calvin's Commentary on Seneca's De Clementia.* Edited by Ford Lewis Battles. Leiden: Brill, 1969.

———. *The Epistle of Paul the Apostle to the Galatians, Philippians, Ephesians, and Colossians.* Translated by T. H. L. Parker. Grand Rapids: Eerdmans, 1964.

———. *Institutes of the Christian Religion.* Translated by Henry Beveridge. 2 vols. London: James Clarke, 1962.

———. *Institutio Christianae Religionis.* Genevæ, 1559.

———. *The Institution of Christian Religion.* Translated by Thomas Norton. London, 1651.

———. *The Institution of the Christian Religion.* Glasgow, 1762.

———. *Tracts and Treatises on the Reformation of the Church.* 7 vols. Translated by Henry Beveridge. Edinburgh, 1844.

Carroll, Lewis. *The Complete Works of Lewis Carroll.* New York: Modern Library, 1936.

Casaubon, Meric. *A Letter of Meric Casaubon . . . Concerning Natural Experimental Philosophie.* Cambridge, 1669.

Chalmers, T. *On the Christian and Civic Economy of Large Towns.* 3 vols. Glasgow, 1821–26.

Chambers, Ephraim. *Cyclopaedia; or, An Universal Dictionary of Arts and Sciences*. 2 vols. London, 1728.

Clark, J. W., and T. Hughes, eds. *The Life and Letters of the Rev. Adam Sedgwick*. 2 vols. Cambridge: Deighton, Bell and Co., 1890.

Clarke, Samuel. *The Works of Samuel Clarke, D.D.* 2 vols. London, 1738.

Coleridge, Samuel Taylor. *The Works of Samuel Taylor Coleridge, Prose and Verse*. Philadelphia: Crissy & Markley, 1846.

Condorcet, Nicolas de. *Sketch for a Historical Picture of the Progress of the Human Mind*. Translated by June Barraclough. New York: Noonday, 1955.

Constantine. "Receipt of Constantine to the Bishops of Numidia." In *The Work of St. Optatus*, translated by O. R. Vassall-Phillips. London: Longmans, Green and Co., 1917.

Cotes, Roger. Preface to Newton's *Principia*. In *Isaac Newton: The Principia*, translated by I. B. Cohen and A. Whitman. Berkeley: University of California Press, 1999.

Culpeper, Thomas. *Morall Discourses and Essayes*. London, 1655.

Delbanco, Andrew, and Alan Heimert, eds. *The Puritans in America: A Narrative Anthology*. Cambridge, MA: Harvard University Press, 1985.

Descartes, René. *The Philosophical Writings of Descartes*. 2 vols. Translated by J. Cottingham, R. Stoothoff, and D. Murdoch. Cambridge: Cambridge University Press, 1984–91.

Disraeli, Isaac. *Calamities and Quarrels of Authors*. London: Warne and Co., 1881.

Downame, George. *A Treatise of Iustification*. London, 1633.

Draper, John William. *History of the Conflict between Religion and Science*. New York: D. Appleton, 1875.

Dungan, David Roberts. *Hermeneutics: A Text Book*. Cincinnati: Standard Publishing Co., 1888.

Duns Scotus, John. *Duns Scotus: Philosophical Writings*. Translated by Allan Wolter. Indianapolis: Hackett, 1987.

———. *Opera Omnia*. Edited by C. Balíc et al. Vatican City: Typis Polyglottis Vaticanis, 1950–.

Edward, Lord Herbert of Cherbury. *De Religione Laici* [1645]. Edited and translated by Harold R. Hutcheson. New Haven: Yale University Press, 1944.

Edwards, Jonathan. *The Works of Jonathan Edwards*. 2 vols. Peabody: Hendrickson, 1998.

Edwards, John. *A Demonstration of the Existence and Providence of God*. London, 1696.

Encyclopaedia Britannica. 3 vols. Edinburgh, 1771.

Epictetus. *The Works of Epictetus. Consisting of His Discourses, in Four Books, The Enchiridion, and Fragments*. Edited by Elizabeth Carter and translated by Thomas Wentworth Higginson. Boston: Little, Brown, and Co., 1865.

Erasmus of Rotterdam. *E Desiderii Erasmi Roterodami Opera Omnia*. 10 vols. Edited by J. Clericus. Leyden, 1703–6.

Euclid. *Thirteen Books of Euclid's Elements*. 3 vols. Translated by Thomas Heath. New York: Dover, 1956.

Eusebius of Caesarea. *Preparation for the Gospel.* Translated by E. H. Gifford. Oxford: Clarendon, 1903.

Eustachius à Sancto Paulo. *Summa Philosophiae Quadripartita.* Paris, 1609.

Evagrius Ponticus. *Scholia In Proverbia.* Paris: Cerf, 1987.

Evelyn, John. *The Diary of John Evelyn.* 3 vols. London: Routledge/Thoemmes, 1996.

Fairholme, George. *A General View of the Geology of Scripture.* London: James Ridgway, 1833.

Ficino, Marsilio. *Opera.* 2 Tom. Basil, 1547.

Flavell, John. *An Exposition of the Assembly's Shorter Catechism.* London, 1688.

Fontenelle, Bernard. *Digression on the Ancients and the Moderns,* in *Conversations with a Lady on the Plurality of Worlds.* Translated by Joseph Glanvill. 4th ed. London, 1719.

Freeland, Edward B. "Buckle, Draper, and a Science of History." *Continental Monthly* 4 (July–September 1863): 610–24.

Fulwell, Ulpian. *The First Parte, of the Eighth Liberall Science: Entituled, Ars Adulandi.* London, 1597.

Galileo Galilei. *The Assayer: A Letter to the Illustrious and Very Reverend Don Virginio Cesarini.* In *Discoveries and Opinions of Galileo,* translated by Stillman Drake, pp. 231–80. New York: Doubleday & Co., 1957.

———. *Dialogue Concerning the Two Chief World Systems—Ptolemaic & Copernican.* Translated by Stillman Drake. Berkeley: University of California Press, 1962.

Galton, Francis. *English Men of Science: Their Nature and Nurture.* London: Macmillan, 1874.

Gerth, H. H., and C. Wright Mill, eds. *From Max Weber: Essays in Sociology.* London: Routledge, 1970.

Ghazālī, Abū Hāmid al-. *The Incoherence of the Philosophers (Tahâfut al-falâsifa).* Translated by P. Marmura. Salt Lake City: Brigham Young University Press, 2002.

Gilby, Anthony, and William Kethe. *The Appellation of Iohn Knoxe from the Cruell and Iniust Sentence Pronounced against Him by the False Bishoppes and Clergie of Scotland, . . . and Comunaltie of the Same Realme.* Geneva, 1558.

Glanvill, Joseph. *Essays on Several Important Subjects in Philosophy and Religion.* London, 1676.

———. *Plus Ultra . . . in an Account of some of the most Remarkable Late Improvements of Practical Useful Learning.* London, 1668.

———. *Scepsis Scientifica.* London, 1665.

Grew, Nehemiah. *Musaeum Regalis Societatis.* London, 1681.

Grosseteste, Robert. *On the Six Days of Creation.* Translated by C. Martin. Oxford: Oxford University Press, 1996.

Hanow, Michael Christoph. *Philosophiae naturalis sive physicae dogmaticae.* Halae Magdeburgicae, 1766.

Harrington, James. *The Prerogative of Popular Government.* London, 1658.

Henslow, John. *Questions on the Subject Matter of Sixteen Lectures in Botany Required for a Pass Examination.* Cambridge: Cambridge University Press, 1851.

Herder, Johann Gottfried. *Philosophical Writings*. Edited by Michael N. Forster. Cambridge: Cambridge University Press, 2002.

Herschel, John. *Preliminary Discourse*. London: Longman, Brown, Green & Longmans, 1851.

Hippocrates. *Ancient Medicine*. Translated by W. H. S. Jones. Cambridge, MA: Harvard University Press, 1923.

Hobbes, Thomas. *The English Works of Thomas Hobbes*. 11 vols. Edited by William Molesworth. London: Bohn, 1839.

———. *Leviathan*. Edited by C. B. Macpherson. Ringwood: Penguin, 1982.

Hodge, Charles. *Systematic Theology* [1872–73]. 3 vols. Grand Rapids: Eerdmans, 1970.

Holyoake, Thomas. *A Large Dictionary in Three Parts*. London, 1676.

Hooke, Robert. *Micrographia*. London, 1665.

Hopkins, Mark. *Evidences of the Christian Religion*. Boston, 1909.

Hugh of Saint Victor. *Didascalicon*. Translated by Jerome Taylor. New York: Columbia University Press, 1991.

Hume, David. *The Natural History of Religion*. Edited by H. E. Root. Stanford: Stanford University Press, 1957.

Huxley, Leonard. *The Life and Letters of Thomas Henry Huxley*. 3 vols. New York: D. Appleton, 1900.

Huxley, Thomas Henry. *Science and the Christian Tradition*. New York: D. Appleton, 1894.

Isidore of Seville. *Etymologiae*. Edited by W. M Lindsay. Oxford: Claendon Press, 1911.

James, William. *The Will to Believe and Other Essays in Popular Philosophy*. New York: Cosimo, 2007.

Jenyns, Soame. *View of the Internal Evidence of the Christian Religion*. London, 1776.

Jevons, W. Stanley. *The Principles of Science: A Treatise on Logic and the Scientific Method*. London: Macmillan, 1874.

Kepler, Johannes. *Gesammelte Werke*. 20 vols. Munich: C. H. Beck, 1937–45.

———. *The Harmony of the World*. Translated and introduced by E. J. Aiton, A. M. Duncan, and J. V. Field. Philadelphia: American Philosophical Society, 1997.

———. *Mysterium Cosmographicum*. Translated by A. M. Duncan. Norwalk, CT: Abaris, 1999.

———. *Selbstzeugnisse*. Edited Franz Hammer and translated by Esther Hammer. Stuttgart-Bad Constatt: Friedrich Frommann Verlag, 1971.

Kirby, William. *On the History, Habits and Instincts of Animals*. Philadelphia: Carey, Lea and Blanchard, 1837.

Lamar, J. S. *The Organon of Scripture; or, The Inductive Method of Biblical Interpretation*. Philadelphia: J. B. Lippincott, 1860.

Lawrence, William. *Lectures on Physiology, Zoology, and the Natural History of Man* [1819]. Salem, 1828.

Lenzer, Gertrud, ed. *Auguste Comte and Positivism: The Essential Writings*. New York: Harper, 1975.

Lessing, Gotthold. *Lessing's Theological Writings*. Edited by Henry Chadwick. Stanford: Stanford University Press, 1957.

Lewis, Charlton T., and Charles Short. *A Latin Dictionary.* Oxford: Clarendon, 1879.

Locke, John. *Essay Concerning Human Understanding* [1690]. Edited by Peter H. Nidditch. Oxford: Clarendon, 1975.

———. *Of the Conduct of the Understanding.* Edited by Thomas Fowler. 5th ed. Oxford: Oxford University Press, 1901.

———. *The Works of John Locke.* 10 vols. 12th ed. London: Thomas Tegg, 1823.

Lockhart, Clinton. *Principles of Interpretation as Recognized Generally by Biblical Scholars, Treated as a Science, Derived Inductively from an Exegesis of Many Passages of Scripture.* 2nd ed. Fort Worth: S. H. Taylor, 1915.

Lund, Eric, ed. *Documents from the History of Lutheranism, 1517-1750.* Minneapolis: Augsburg Fortress, 2002.

Luther, Martin. *The Complete Sermons of Martin Luther.* 7 vols. Edited by John N. Lenker. Grand Rapids: Baker Books, 2000.

———. *Luther's Works.* 55 vols. Edited by J. Pelikan and H. Lehman. St. Louis: Concordia, 1955-75.

———. *Three Treatises.* Translated by W. A. Lambert. Philadelphia: Fortress Press, 1970.

Macaulay, Thomas Babington. *Critical, Historical and Miscellaneous Essays.* 6 vols. New York: Hurd and Houghton, 1860.

———. *The History of England from the Accession of James II.* 2 vols. London: Longman, Brown, Green and Longmans, 1849.

———. "Lord Bacon," *Edinburgh Review,* July 1837, 1-104.

Mather, Cotton. *American Tears upon the Ruines of the Greek Churches.* Boston, 1701.

McCulloch, J. R., ed. *Early English Tracts on Commerce.* Cambridge: Cambridge University Press, 1952.

Mersenne, Marin. *La vérité des sciences contre les septiques ou Pyrrhoniens.* Paris, 1625.

Newman, John Henry. *Essay on the Development of Christian Doctrine.* London: James Toovey, 1845.

———. *Fifteen Sermons Preached before the University of Oxford.* New York: Longmans, Green & Co., 1918-20.

Newton, Isaac. *Isaac Newton: The Principia.* Translated by I. B. Cohen and A. Whitman. Berkeley: University of California Press, 1999.

Origen. *Contra Celsum.* Edited and translated by Henry Chadwick. Cambridge: Cambridge University Press, 1965.

Orwell, George. *Nineteen Eighty-Four: A Novel.* New York: Plume, 2003.

Owen, John. *Works of John Owen, D.D.* 24 vols. Edited by W. H. Gould. Edinburgh: T. & T. Clark, 1850-53.

Paley, William. *Paley's Works: Consisting of Evidences of Christianity; More and Political Philosophy; Natural Theology, and Horae Pauline.* London: Bohn, 1849.

———. *View of the Evidences of Christianity.* 2 vols. London, 1794.

Pascal, Blaise. *Pascal: Œuvres complètes.* Edited by Louis Lafuma. Paris: Éditions du Seuil, 1963.

———. *Thoughts, Letters, Minor Works.* New York: Cosimo Books, 2009.

Pattison, Mark. *Essays by the Late Mark Pattison.* Oxford: Clarendon, 1889.

Pepys, Samuel. Vol. 5 of *The Diary of Samuel Pepys*. Edited by R. Latham and W. Matthews. London: Bell, 1971.

Perkins, William. *A Golden Chaine*. London, 1600.

Physiologus: A Medieval Book of Nature Lore. Translated by Michael Curley. Chicago: University of Chicago Press, 1979.

[Plattes, Gabriel]. *A Description of the Famous Kingdome of Macaria*. London, 1641.

Plato. *The Collected Dialogues of Plato*. Edited by Edith Hamilton and Huntington Cairns. Princeton: Princeton University Press, 2005.

Pluche, Noël-Antoine. *Spectacle de la Nature; or, Nature Display'd*. 7 vols. London, 1770.

Proast, Jonas. *The Argument of the "Letter Concerning Toleration" Considered and Answered*. Oxford, 1690.

Pufendorf, Samuel. *The Law of Nature and of Nations*. Translated by C. H. Oldfather and W. A. Oldfather. Oxford: Clarendon, 1934.

———. *Of the Nature and Qualification of Religion in Reference to Civil Society*. Translated by Jodocus Crull and edited by Simone Zurbuchen. Indianapolis: Liberty Fund, 2001.

———. *The Whole Duty of Man According to the Law of Nature*. Translated by Andrew Tooke and edited by Ian Hunter and David Saunders. With *Two Discourses and a Commentary by Jean Barbeyrac*. Translated by David Saunders. Indianapolis: Liberty Fund, 2003.

Quasten, Johannes. *Patrology*. 4 vols. Allen: Christian Classics, 1986.

Ray, John. *The Wisdom of God Manifested in the Works of Creation*. London, 1691.

Ramon Sibiuda [Raymond of Sabunde]. *Theologia naturalis seu liber creaturarum*. Edited by F. Stegmüller. Stuttgart-Bad Cannstatt: Frommann, 1966.

Ray, John, and Francis Willoughby. *The Ornithology of Francis Willughby*. London, 1678.

Recorde, Robert. *The Urinal of Physick*. London, 1651.

Renan, Ernest. "What Is a Nation?" In *Nation and Narration*, edited by Homi K. Bhabha and translated by Martin Thom, pp. 8–22. London: Routledge, 1990.

Richard of Saint Victor. *The Twelve Patriarchs, the Mystical Ark, Book Three of the Trinity*. Translated by Grover A. Zinn. New York: Paulist Press, 1979.

Richardson, Cyril C., ed. *Early Christian Fathers*. New York: Touchstone, 1996.

[Robertson, James Burton]. "Science and Revealed Religion." *Dublin Review*, April 1837, 293–329.

Roget, Peter. *Animal and Vegetable Physiology Considered with Reference to Natural Theology*. 2 vols. Philadelphia: Carey, Lea & Blanchard, 1836.

Securis, John. *Detection and querimonie of the daily enormities and abuses comitted in physic*. London, 1566.

Sedgwick, Adam. "Objections to Mr Darwin's *Origin of Species*." *The Spectator*, 24 March 1860.

[Sedgwick, Adam]. Review of *Vestiges of the Natural History of Creation*. *Edinburgh Review* 82 (July 1845): 1–85.

Seneca. *Natural Questions*. Books 1–3. Edited by Jeffrey Henderson. Translated by Thomas H. Corcoran. Cambridge, MA: Harvard University Press, 1971.

Sergeant, John. *The Method to Science*. London, 1696.

Shadwell, Thomas. *The Virtuoso*. London, 1676.

Simplicius. *In Physica*. Vol. 9, *Commentaria in Aristotelem Graeca*, edited by H. Diels. Berlin: Reimer, 1882.

South, Robert. *Twelve sermons upon several subjects and occasions*. Vol. 3. London, 1693.

———. *Twelve Sermons Upon Several Subjects and Occasions*. 3 vols. London, 1697–98.

Sprat, Thomas. *History of the Royal Society, for the Improving of Natural Knowledge*. London, 1667.

———. *Sermons Preached upon Several Occasions*. London, 1697.

Stubbe, Henry. *Campanella Revived*. London, 1670.

———. *A Censure upon Certain Passages Contained in the History of the Royall Society*. 2nd ed. Oxford, 1671.

———. *Plus Ultra Reduced to a Non-Plus*. London, 1670.

Suárez, Francisco. *On Efficient Causation*. Translated by Alfred Freddoso. New Haven: Yale University Press, 1994.

Swift, Jonathan. *Travels into Several Remote Regions of the World*. London, 1726.

———. *The Works of the Rev. Jonathan Swift, D.D.* 24 vols. London, 1803.

Taylor, Jeremy. *Symbolon theologikon*. London, 1674.

Thomas Aquinas. *Commentary on the Sentences of Peter Lombard*. Toronto: Pontifical Institute, 1997.

———. *Compendium of Theology*. Translated by C. Vollert. Saint Louis: Herder, 1948.

———. *Division and Methods of the Sciences: Questions V and VI of his Commentary on the "De Trinitate" of Boethius*. Translated by Armand Maurier. Toronto: Pontifical Institute of Medieval Studies, 1953.

———. *On the Power of God*. Translated by the English Dominican Friars. 3 vols. London: Burns, Oates & Washbourne, 1932–34.

———. *Summa theologiae*. 60 vols. London: Blackfriars, 1964–76.

Thomasius, Christian. *On the Right of Protestant Princes Regarding Heretics* [1697]. In Ian Hunter, *The Secularization of the Confessional State: The Political Thought of Christian Thomasius*, pp. 168–206. Cambridge: Cambridge University Press, 2007.

Thomson, William. *Outline of the Laws of Thought*. London: Pickering, 1842.

Toledo, Francisco. *Commentaria una cum quæstionibus in universam Aristotelis logicam Post. anal.* Rome, 1572.

Topsel, Edward. *The Historie of Foure-Footed Beastes*. London, 1607.

Vaughan, William. *The Golden-Groue Moralized in Three Bookes: A Worke Very Necessary for All Such, as Would Know How To Gouerne Themselues, Their Houses, Or Their Countrey*. London, 1600.

Vincent of Beauvais. *Speculum quadruplex naturale, doctrinale, morale, historiale*. 4 vols. Douai, 1624.

Wallis, John. *Opera mathematica*. 3 vols. Oxford, 1699.

Ward, W. G. "Science, Prayer, Free Will, and Miracles." *Dublin Review* 8, no. 16 (April 1867): 255–98.

Wayland, Francis. *The Elements of Moral Science*. New York: Cooke and Co., 1835.

Weber, Max. *From Max Weber: Essays in Sociology*. Edited by H. H. Gert and C. Wright
    Mill. London: Routledge, 1970.
———. *The Protestant Ethic and the Spirit of Capitalism*. Translated by Talcott Parsons.
    London: Routledge, 2001.
Wesley, John. *The Works of the Rev. John Wesley*. 7 vols. Edited by John Emory. 3rd. ed.
    New York: Carlton and Porter, 1856.
Westcott, Brooke Foss. *The Gospel of Life*. London: Macmillan, 1892.
Whewell, William. *History of the Inductive Sciences, from the Earliest to the Present Time*.
    3 vols. London, 1837.
———. *Indications of the Creator*. 2nd ed. London: John Parker, 1846.
———. *On the Principles of the English University Education*. London, 1837.
———. *The Philosophy of the Inductive Sciences: Founded upon their History*. 2 vols. Lon-
    don: John Parker, 1847.
[Whewell, William]. "Mrs Sommerville on the Connexion of the Physical Sciences."
    Quarterly Review 51 (March and June 1834): 54–68.
Whiston, William. *A New Theory of the Earth*. London, 1696.
White, Andrew Dickson. *History of the Warfare of Science with Theology in Christendom*.
    2 vols. New York: D. Appleton, 1896.
White, G. Frederick. Review of *Darwinism and Other Essays*, by John Fiske. *Bibliotheca
    Sacra* 36 (1879): 784.
William of Auvergne. *The Universe of Creatures*. Translated by Roland J. Teske. Milwau-
    kee: Marquette University Press, 1998.
Wiseman, Nicholas. *Lectures on the Connexion between Science and Revealed Reli-
    gion*. 2 vols. London: Booker, 1836.
Woods, Leonard. *Letters to Unitarians and Reply to Dr. Ware*. Andover: Mark Newton,
    1822.
Zwingli, Ulrich. *The Latin Works of Huldreich Zwingli*. 3 vols. Edited by Clarence N.
    Heller. Philadelphia: Heidelberg Press, 1929.

Secondary Sources

Ackeran, Gerald van. *Sacra Doctrina*. Rome: Catholic Book Agency, 1952.
Aertsen, Jan A., Kent Emery Jr., and Andreas Speer, eds. *Nach der Verurteilung von 1277.
    Philosophie und Theologie an der Universität von Paris im letzten Viertel des 13. Jahrhun-
    derts. Studien und Texte*. Berlin: de Gruyter, 2001.
Algra, Keimpe. "The Beginnings of Cosmology." In Long, *The Cambridge Companion to
    Ancient Greek Philosophy*, pp. 45–65.
———. "Stoic Theology." In *The Cambridge Companion to the Stoics*, edited by Brad
    Inwood, pp. 153–78. Cambridge: Cambridge University Press, 2003.
Allan, G. A. T. *Christ's Hospital*. Shepperton: Town and County, 1984.
Allen, James. *Inference from Signs: Ancient Debates about the Nature of Evidence*. Oxford:
    Oxford University Press, 2001.
Almond, Philip C. *The British Discovery of Buddhism*. Cambridge: Cambridge University
    Press, 1988.

Alston, William. *Perceiving God: The Epistemology of Religious Experience.* Ithaca: Cornell University Press, 1991.

Althaus, Paul. *The Theology of Martin Luther.* Philadelphia: Fortress, 1966.

Anstey, Peter. "Boyle on Seminal Priciples." *Studies in History and Philosophy of Science, Part C,* 33 (2002): 597–630.

———. "Locke on Method in Natural Philosophy." In *The Philosophy of John Locke: New Perspectives,* edited by Peter Anstey, pp. 26–42. London: Routledge, 2003.

Armitage, David. "What's the Big Idea? Intellectual History and the *Longue Durée.*" *History of European Ideas* 38 (2012): 493–507.

Armstrong, Patrick. *The English Parson-Naturalist: A Companionship between Science and Religion.* Leominster, Herefordshire: Gracewing, 2000.

Atran, Scott. *In Gods We Trust: The Evolutionary Landscape of Religion.* Oxford: Oxford University Press, 2002.

Attfield, Robin. *The Ethics of Environmental Concern.* Oxford: Blackwell, 1983.

Austen, J. L. *Doing Things with Words.* Oxford: Oxford University Press, 1962.

Barbour, Ian. *Religion and Science: Historical and Contemporary Issues.* San Francisco: Harper & Row, 1997.

———. *Western Man and Environmental Ethics: Attitudes towards Nature and Technology.* Reading, MA: Addison-Wesley, 1973.

Barton, Ruth. "'An Influential Set of Chaps': The X-Club and Royal Society Politics, 1864–85." *British Journal for the History of Science* 23 (1990): 53–81.

Beard, Mary, John North, and Simon Price. *The Religions of Rome.* 2 vols. Cambridge: Cambridge University Press, 1998.

Bebbington, D. W. *Evangelicalism in Modern Britain: A History from the 1730s to the 1980s.* London: Unwin Hyman, 1989.

Beck, Adam H., and Annett Yoshiko Reed, eds. *The Ways That Never Parted: Jews and Christians in Late Antiquity and the Early Middle Ages.* Minneapolis: Fortress Press, 2007.

Benedict, Barbara M. *Curiosity: A Cultural History of Early Modern Inquiry.* Chicago: Chicago University Press, 2001.

Bennett, Tony. *The Birth of the Museum: History, Theory, Politics.* London: Routledge, 1995.

Berchman, Robert M. *From Philo to Origen: Middle Platonism in Transition.* Chico, CA: Scholars Press, 1984.

Berdyaev, Nikolai. *Slavery and Freedom.* New York: Charles Scribner's Sons, 1944.

Berkel, Klaas van, and Arie Vanderjagt, eds. *The Book of Nature in Antiquity and the Middle Ages.* Louvain: Peeters, 2005.

———, eds. *The Book of Nature in Early Modern and Modern History.* Louvain: Peeters, 2006.

Bevir, Mark. "*Begriffsgeschichte.*" *History and Theory* 39 (2000): 273–84.

———. "The Errors of Linguistic Contextualism," *History and Theory* 31 (1992): 276–98.

———. *The Logic of the History of Ideas.* Cambridge: Cambridge University Press, 1999.

Beyer, Peter. "The Modern Emergence of Religions and a Global Social System for Religion." *International Sociology* 13 (1998): 151–72.

Bickerman, Elias J. "The Name of Christians." *Harvard Theological Review* 42 (1949): 109–24.

Biller, Peter. "Words and the Medieval Notions of 'Religion.'" *Journal of Ecclesiastical History* 36 (1985): 351–69.

Blackwell, Constance. "Neo-Platonic Modes of Concordism versus Definitions of Difference." In *Laus Platonici Philosophi: Marsilio Ficino and His Influence*, edited by Stephen Clucas, Peter J. Forshaw, and Valery Rees, pp. 317–42. Leiden: Brill, 2011.

Blair, Ann. "Natural Philosophy." In *The Cambridge History of Science*, vol. 3, *Early Modern Science*, edited by Katherine Park and Lorraine Daston, pp. 365–406. Cambridge: Cambridge University Press, 2007.

———. *Too Much to Know: Managing Scholarly Information before the Modern Age*. New Haven: Yale University Press, 2010.

Bloor, David. "Epistemic Grace: Antirelativism as Theology in Disguise." *Common Knowledge* 12 (2007): 250–80.

———. "Rationalism, Supernaturalism, and the Sociology of Knowledge." In *Scientific Knowledge Socialized: Selected Proceedings of the 5th Joint International Conference on the History and Philosophy of Science Organized by the IUHPS, Veszprém, 1984*, ed. I. Hronszk, M. Fehér, and B. Dajka, pp. 59–74. Dordrecht: Kluwer, 1988.

Blowers, Paul M. "'Entering this Sublime and Blessed Amphitheatre': Contemplation of Nature and Interpretation of the Bible in the Patristic Period." In van der Meer and Mandelbrote, *Nature and Scripture in the Abrahamic Traditions*, pp. 147–76.

Blumenberg, Hans. *Work on Myth*. Translated by Robert M. Wallace. Cambridge, MA: MIT Press, 1985.

Bossy, John. "Some Elementary Forms of Durkheim." *Past and Present* 95 (1982): 3–18.

Boswell, James. *Life of Johnson*. 4 vols. Edited by George B. Hill and Lawrence Powell. Oxford: Clarendon, 1994.

Boyarin, Daniel. *Border Lines: The Partition of Judeao-Christianity*. Philadelphia: University of Pennsylvania Press, 2004.

———. "The Christian Invention of Judaism." In *Religion: Beyond a Concept*, edited by Hent de Vries, pp. 150–77. New York: Fordham University Press, 2008.

Boyer, Pascal. *The Naturalness of Religious Ideas: A Cognitive Theory of Religion*. Berkeley: University of California Press, 1994.

———. *Religion Explained: The Evolutionary Origins of Religious Thought*. New York: Basic Books, 2001.

Bozeman, Theodore. *Protestants in an Age of Science: The Baconian Ideal and Antebellum American Religious Thought*. Chapel Hill: University of North Carolina Press, 1977.

Brady, Jules. "Saint Augustine's Theory of Seminal Reasons." *New Scholasticism* 38 (1964): 141–58.

Brague, Rémi. *The Legend of the Middle Ages: Philosophical Explorations of Medieval Christianity, Judaism, and Islam*. Translated by Lydia G. Cochrane. Chicago: University of Chicago Press, 2009.

———. *The Wisdom of the World: The Human Experience of the Universe in Western Thought*. Translated by Teresa Fagan. Chicago: University of Chicago Press, 2003.

Brickhouse, Thomas C., and Nicholas D. Smith. *Plato's Socrates*. Oxford: Oxford University Press, 1994.

Brisson, Luc. *How Philosophers Saved Myths*. Chicago: University of Chicago Press, 2004.

Broadie, Sarah. "Rational Theology." In Long, *The Cambridge Companion to Ancient Greek Philosophy*, pp. 205–24.

Brockman, John, ed. *What Is Your Dangerous Idea?* New York: Harper, 2007.

Brooke, Christopher. *A History of the University of Cambridge*. Vol. 4, *1870–1990*. Cambridge: Cambridge University Press, 1988.

Brooke, John Hedley, and Geoffrey Cantor. *Reconstructing Nature: The Engagement of Science and Religion*. Edinburgh: T. & T. Clark, 1998.

Brooke, John Hedley. "Natural Theology and the Plurality of Worlds: Observations on the Brewster-Whewell Debate." *Annals of Science* 34 (1977): 221–86.

Brooke, John Hedley, and Ronald Numbers, eds. *Science and Religion around the World: Historical Perspectives*. Oxford: Oxford University Press, 2011.

Brown, Stephen. "Key Terms in Medieval Theological Vocabulary." In *Méthodes et instruments du travail intellectuel au moyen âge*, edited by Olga Weijers, pp. 82–97. Turnhout: Brepols, 1990.

Browne, Thomas. *Pseudodoxia Epidemica*. Edited by Robin Robbins. 2 vols. Oxford: Clarendon, 1981.

Buckley, Michael J. *At the Origins of Modern Atheism*. New Haven: Yale University Press, 1987.

Buell, Denise Kimber. "Race and Universalism in Early Christianity." *Journal of Early Christian Studies* 10 (2002): 429–68.

———. *Why This New Race? Ethnic Reasoning within Early Christianity*. New York: Columbia University Press, 2005.

Bultmann, Rudolf. *Theology of the New Testament*. 2 vols. London: SCM, 1965.

Bunch, Bryan, and Alexander Hellemans. *The History of Science and Technology*. Boston: Houghton Mifflin, 2004.

Bunge, Mario. *Causality and Modern Science*. New York: Dover, 1979.

Bunsen, Christian. *Hippolytus and his Age*. 4 vols. London: Longman, Brown, Green and Longmans, 1852.

Burnet, John. *Early Greek Philosophy*. 3rd ed. London: A & C. Black, 1920.

Bury, J. B. *The Idea of Progress: An Inquiry into Its Origins and Growth*. London: Macmillan, 1932.

Butterfield, Herbert. *The Origins of Modern Science: 1300–1800*. 2nd ed. New York: Macmillan, 1962.

———. *The Whig Interpretation of History*. New York: W. W. Norton, 1965.

Buxton, Richard ed. *From Myth to Reason? Studies in the Development of Greek Thought*. Oxford: Oxford University Press, 1999.

Calhoun, Craig, Mark Juergensmeyer, and Jonathan VanAntwerpen, eds. *Rethinking Secularism*. Oxford: Oxford University Press, 2011.

Caneva, K. "From Galvanism to Electrodynamics: The Transformation of German

Physics and its Social Context." *Historical Studies in the Physical Sciences* 9 (1978): 63–159.

Canevet, Mariette. *Grégoire de Nysse et l'herméneutique biblique*. Paris: Études Augustiniennes, 1983.

Cantor, Geoffrey, and Chris Kenny. "Barbour's Fourfold Way: Problems with His Taxonomy of Science-Religion Relationships." *Zygon* 36 (2001): 765–81.

Carr, Nicholas. *The Shallows: How the Internet Is Changing the Way We Read, Think and Remember*. London: Atlantic Books, 2010.

Carraud, Vincent. *Causa sive Ratio: La raison de la cause, de Suarez à Leibniz*. Paris: Presses Universitaires de France, 2002.

Cartwright, Nancy. *The Dappled World: A Study of the Boundaries of Science*. Cambridge: Cambridge University Press, 1999.

Cassirer, Ernst. *The Philosophy of Symbolic Forms*. 2 vols. Translated Ralph Manheim. New Haven: Yale University Press, 1955.

Cavanaugh, William T. *The Myth of Religious Violence: Secular Ideology and the Roots of Modern Conflict*. Oxford: Oxford University Press, 2009.

Chaisson, Eric. *Epic of Evolution: Seven Ages of the Cosmos*. New York: Columbia University Press, 2006.

Charbonneau-Lassay, Louis. *The Bestiary of Christ*. New York: Arcana, 1972.

Christian, David. *Maps of Time: An Introduction to Big History*. Berkeley: University of California Press, 2004.

Clay, Diskin. *Paradosis and Survival: Three Chapters in the History of Epicurean Philosophy*. Ann Arbor: University of Michigan Press, 1998.

Cohen, J. D. *The Beginnings of Jewishness: Boundaries, Varieties, Uncertainties*. Berkeley: University of California Press, 1999.

Cohen, Jeremy. *"Be Fertile and Increase, Fill the Earth and Master It": The Ancient and Medieval Career of a Biblical Text*. Ithaca: Cornell University Press, 1989.

Colish, Marcia L. *The Stoic Tradition from Antiquity to the Early Middle Ages*. 2 vols. Leiden: Brill, 1985.

Collins, Arthur. *Symbolism of Animals and Birds Represented in English Church Architecture*. London: Pitman and Sons, 1913.

Condren, Conal, Stephen Gaukroger, and Ian Hunter, eds. *The Philosopher in Early Modern Europe: The Nature of a Contested Identity*. Cambridge: Cambridge University Press, 2006.

Corneanu, Sorana. *Regimens of the Mind: Boyle, Locke, and the Early Modern Cultura Animi Tradition*. Chicago: University of Chicago Press, 2011.

Cornford, F. M. *From Philosophy to Religion: A Study in the Origins of Western Speculation* [1912]. Princeton: Princeton University Press, 1991.

Corsi, Pietro. "Biologie." In *Lamarck, philosophe de la nature*, edited by Pietro Corsi, Jean Gayon, Gabriel Gohau, and Stéphane Tirard, pp. 37–64. Paris: Presses Universitaires de France, 2006.

Cottingham, John. "Descartes as Sage: Spiritual *askesis* in Cartesian Philosophy." In Condren, Gaukroger, and Hunter, *The Philosopher in Early Modern Europe*, 182–201.

————. *Philosophy and the Good Life: Reason and the Passions in Greek, Cartesian and Psychoanalytical Ethics.* Cambridge: Cambridge University Press, 1998.

Cox, Patricia. "Origen and the Bestial Soul." *Vigiliae Christianae* 36 (1982): 115–40.

Crane, Ronald S. "Anglican Apologetics and the Idea of Progress." *Modern Philology* 31 (1934): 273–306.

Crisp, Roger, and Michael Slote, eds. *Virtue Ethics.* Oxford: Oxford University Press, 1997.

Crombie, A. C. *Augustine to Galileo.* 2nd ed. Harmondsworth: Penguin, 1969.

Cross, Richard. *Duns Scotus.* Oxford: Oxford University Press, 1999.

————. "Duns Scotus and Suárez at the Origins of Modernity." In *Deconstructing Radical Orthodoxy: Postmodern Theology, Rhetoric and Truth,* edited by J. Hankey and Douglas Hedley, pp. 65–80. Aldershot: Ashgate, 2005.

Crosson, Frederick J. "'Fides' and 'Credere': W. C. Smith on Aquinas." *Journal of Religion* 65 (1985): 399–412.

Cunningham, Andrew. "Getting the Game Right: Some Plain Words on the Identity and Invention of Science." *Studies in History and Philosophy of Science* 19 (1988): 365–89.

————. "How the *Principia* Got Its Name; or, Taking Natural Philosophy Seriously." *History of Science* 28 (1991): 377–92.

Curtius, E. R. *European Literature and the Latin Middle Ages.* Translated by W. R. Trask. London: Routledge and Kegan Paul, 1953.

Daniélou, Jean. *Platonisme et théologie mystique.* 2nd ed. Paris: Editions Montaigne, 1953.

Daston, Lorraine. "Attention and the Values of Nature in the Enlightenment." In *The Moral Authority of Nature,* edited by Lorraine Daston and Fernando Vidal, pp. 100–26. Chicago: University of Chicago Press, 2004.

Davies, Paul. *The Goldilocks Enigma: Why Is the Universe Just Right for Life?* London: Penguin, 2007.

Dawkins, Richard. *The God Delusion.* Boston: Houghton Mifflin, 2006.

————. "Is Science a Religion?" *The Humanist,* January–February 1977.

Dear, Peter. *The Intelligibility of Nature: How Science Makes Sense of the World.* Chicago: University of Chicago Press, 2006.

————. "Method and the Study of Nature." In *The Cambridge History of Seventeenth-Century Philosophy,* edited by D. Garber and M. Ayers, 1:147–77. 2 vols. Cambridge: Cambridge University Press, 1998.

————. "Religion, Science, and Natural Philosophy: Thoughts on Cunningham's Thesis." *Studies in History and Philosophy of Science* 32 (2001): 377–86.

————. "What Is the History of Science the History *Of*? Early Modern Roots of the Ideology of Modern Science." *Isis* 96 (2005): 390–406.

Deferrari, Roy J., M. Inviolata Barry, and Ignatius McGuiness. *A Lexicon of St. Thomas Aquinas.* Baltimore: John Lucas, 1948.

Delumeau, Jean. *Catholicism between Luther and Voltaire.* London: Burns and Oates, 1977.

Deming, David. *Science and Technology in World History.* Vol. 2, *Early Christianity, the Rise of Islam and the Middle Ages.* Jefferson, NC: McFarland, 2010.

Dennett, Daniel. *Breaking the Spell: Religion as a Natural Phenomenon.* Harmondsworth: Penguin, 2006.

Desmond, Adrian. *Huxley: From Devil's Disciple to Evolution's High Priest.* 2 vols. London: Michael Joseph, 1994–97.

Desmond, Adrian, and James R. Moore. *Darwin.* London: Michael Joseph, 1991.

Despland, Michel. *La religion en occident: Évolution des idées et du vécu.* Montreal: Fides, 1979.

Detienne, Marcel. *The Masters of Truth in Archaic Greece.* Translated by Janet Lloyd. New York: Zone, 1999.

Deutsch, Karl. *Nationalism and Its Alternatives.* New York: Knopf, 1969.

Diels, Hermann, and Walther Kratz. *Die Fragmente der Vorsokratiker.* 3 vols. 11th ed. Zurich: Wiedmann, 1964.

Dillon, John. *The Heirs of Plato: A Study of the Old Academy, 347–274 BC.* Oxford: Oxford University Press, 2003.

Dilworth, Craig. *The Metaphysics of Science.* Dordrecht: Kluwer, 1995.

Dodds, E. R. *Pagan and Christian in an Age of Anxiety: Some Aspects of Religious Experience from Marcus Aurelius to Constantine.* Cambridge: Cambridge University Press, 1965.

Dodds, Michael J. *Unlocking Divine Action: Contemporary Science and Thomas Aquinas.* Washington, DC: Catholic University of America Press, 2012.

Drozdek, Adam. *Greek Philosophers as Theologians: The Divine Arche.* Aldershot: Ashgate, 2007.

Dupré, John. *The Disorder of Things: Metaphysical Foundations of the Disunity of Science.* Cambridge, MA: Harvard University Press, 1993.

Dupré, Louis. *Passage to Modernity.* New Haven: Yale University Press, 1993.

Ebeling, Gerhard. *Luther: An Introduction to His Thought.* London: Collins, 1970.

Edwards, Karen L. *Milton and the Natural World: Science and Poetry in Paradise Lost.* Cambridge: Cambridge University Press, 1999.

Edwards, Mark. *Neoplatonic Saints.* Liverpool: Liverpool University Press, 2000.

Ehrenpreis, Stefan, and Ute Lotz-Heumann. *Reformation und konfessionelles Zeitalter.* Darmstadt: Wissenschaftliche Buchgesellschaft, 2002.

Elders, Leo. "Justification des 'cinq voies.'" *Revue Thomiste* 61 (1961): 207–25.

———. *The Philosophical Theology of Aquinas.* Leiden: Brill, 1990.

Engelmann, Edward. "Aristotelian Teleology, Presocratic Hylozoism, and 20th Century Interpretations." *American Catholic Philosophical Quarterly* 64 (1990): 297–312.

Feierman, Steven, and John M. Janzen, "African Religions." In Brooke and Numbers, *Science and Religion,* pp. 229–51.

Feil, Ernst. *Religio: Die Geschichte eines neuzeitlichen Grundbegriffs vom Frühchristentum bis zur Reformation.* Göttingen: Vandenhoeck and Ruprecht, 1986.

Ferguson, Niall. *Civilization: The West and the Rest.* London: Penguin Books, 2012.

Fisch, H. "The Scientist as Priest: A Note on Robert Boyle's Natural Theology." *Isis* 44 (1953): 252–65.

Foster, M. B. "The Christian Doctrine of Creation and the Rise of Modern Natural Science." *Mind* 43 (1934): 446–68.

Foucault, Michel. *Discipline and Punish: The Birth of the Prison.* Translated by Alan Sheridan. New York: Vintage, 1979.

———. *The History of Sexuality.* Vol. 3, *The Care of the Self.* Translated by Robert Hurley. New York: Vintage, 1986.

Freddoso, Alfred. "God's General Concurrence with Secondary Causes: Why Conservation Is Not Enough." *Philosophical Perspectives* 5 (1991): 553–85.

———. "Medieval Aristotelianism and the Case against Secondary Causation in Nature." In *Divine and Human Action: Essays in the Metaphysics of Theism*, edited by Thomas V. Morris, pp. 74–118. Ithaca: Cornell University Press, 1988.

Freedman, William. "Swift's Struldbrugs, Progress, and the Analogy of History." *Studies in English Literature, 1500–1900* 35 (1995): 457–72.

Freeman, Charles. *The Closing of the Western Mind: The Rise of Faith and the Fall of Reason.* New York: Vintage, 2005.

Fukuyama, Francis. *The End of History and the Last Man.* New York: Free Press, 1992.

Funkenstein, Amos. *Theology and the Scientific Imagination: From the Middle Ages to the Seventeenth Century.* Princeton: Princeton University Press, 1986.

Galison, Peter, and David J. Stump, eds. *The Disunity of Science: Boundaries, Contexts and Power.* Stanford: Stanford University Press, 1996.

Gaukroger, Stephen. *The Collapse of Mechanism and the Rise of Sensibility: Science and the Shaping of Modernity, 1680–1760.* Oxford: Oxford University Press, 2010.

———. *The Emergence of a Scientific Culture: Science and the Shaping of Modernity, 1210–1685.* Oxford: Oxford University Press, 2006.

———. *Francis Bacon and the Transformation of Early Modern Natural Philosophy.* Cambridge: Cambridge University Press, 2001.

———. "Science, Religion and Modernity." *Critical Quarterly* 47 (2005): 1–31.

Gay, Peter. *The Enlightenment: An Interpretation.* New York: Vintage, 1968.

Gillespie, Michael Allen. *The Theological Origins of Modernity.* Chicago: University of Chicago Press, 2008.

Gilson, Étienne. *Index Scholastico-Cartésien.* Paris, 1912; reprinted New York: Franklin, 1964.

Ginther, James R. *Master of the Sacred Page: A Study of the Theology of Robert Grosseteste, ca. 1229/30–1235.* Aldershot: Ashgate, 2004.

Glacken, Clarence. *Traces on the Rhodian Shore: Nature and Culture in Western Thought from Ancient Times to the End of the Eighteenth Century.* Berkeley: University of California Press, 1973.

Gorski, Philip S. *The Disciplinary Revolution: Calvinism and the Rise of the State in Early Modern Europe.* Chicago: University of Chicago Press, 2003.

Gowan, Donald, and Millard Shumaker. *Subduing the Earth: An Exchange of Views.* Kingston: United Church of Canada, 1980.

Grant, Edward. "The Condemnation of 1277, God's Absolute Power, and Physical
    Thought in the Late Middle Ages." *Viator* 10 (1979): 211-44.
———. "God and Natural Philosophy: The Late Middle Ages and Sir Isaac Newton."
    Early Science and Medicine 6 (2000): 279-98.
Grant, R. M. "Charges of 'Immorality' against Various Religious Groups in Antiquity."
    In *Studies in Gnosticism and Hellenistic Religions*, edited by R. van den Broek and
    J. Vermaseren, pp. 161-70. Leiden: Brill, 1981.
Grayling, A. C. *Towards the Light: The Struggles for Liberty and Rights that Made the Mod-
    ern West*. London: Bloomsbury, 2007.
Green, Ian. *The Christian's ABC: Catechisms and Catechizing in England c. 1530-1740*.
    Oxford: Clarendon, 1996.
Gregg, Robert R. *Consolation Philosophy: Greek and Christian Paideia in Basil and the Two
    Gregories*. Cambridge, MA: Philadelphia Patristic Foundation, 1975.
Gregory, Brad S., *The Unintended Reformation*. Cambridge, MA: Harvard University
    Press, 2012.
Gulley, N. *The Philosophy of Socrates*. London: Macmillan, 1968.
Guthrie, W. K. C. *History of Greek Philosophy*. 6 vols. Cambridge: Cambridge University
    Press, 1979.
Hadot, Pierre. *La Citadelle intérieure: Introduction aux Pensées de Marc Aurèle*. Paris:
    Fayard, 1992.
———. "Les divisions des parties de la philosophie." *Museum Helveticum* 36 (1978):
    201-23.
———. *Philosophy as a Way of Life*. Translated by Arnold I. Davidson. Oxford: Blackwell,
    1995.
———. *What Is Ancient Philosophy?* Translated by Michael Chase. Cambridge, MA: Har-
    vard University Press, 2002.
———. *Wittgenstein et les limites du langage*. Paris: Vrin, 2004.
Hall, H. R. *The Scientific Revolution, 1500-1800*. 2nd ed. London: Longmans, 1962.
Hampsher-Monk, Iain, Karin Tilmans, and Frank van Vree, eds. *History of Concepts:
    Comparative Perspectives*. Amsterdam: Amsterdam University Press, 1998.
Harkness, Deborah. *The Jewell House: Elizabethan London and the Scientific Revolution*.
    New Haven: Yale University Press, 2007.
Harland, Philip. *Dynamics of Identity in the World of the Early Christians*. London:
    T. & T. Clark, 2009.
Harnack, Adolf. *What Is Christianity?* Translated by Thomas Bailey Saunders. 2nd ed.
    New York: G. P. Putnam's Sons, 1908.
Harris, Roy. *Semantics of Science*. London: Continuum, 2005.
Harris, Sam. "Science Must Destroy Religion." In *What Is Your Dangerous Idea?*, ed. John
    Brockman, pp. 148-51. New York: Harper, 2007.
———. *The Moral Landscape*. London: Transworld, 2010.
Harrison, Peter. *The Bible, Protestantism and the Rise of Science*. Cambridge: Cambridge
    University Press, 2001.
———. "The Cultural Authority of Natural History in Early Modern Europe." In *Biology*

and Ideology from Descartes to Dawkins, edited by Denis Alexander and Ronald Numbers, pp. 11–35. Chicago: University of Chicago Press, 2010.

———. "Curiosity, Forbidden Knowledge, and the Reformation of Natural Philosophy in Early-Modern England." *Isis* 92 (2001): 265–90.

———. "The Development of the Concept of Laws of Nature." In *Creation: Law and Probability,* edited by Fraser Watts, pp. 13–36. Aldershot: Ashgate, 2008.

———. "Experimental Religion and Experimental Science in Early Modern England." *Intellectual History Review* 21 (2011): 413–33.

———. *The Fall of Man and the Foundations of Science.* Cambridge: Cambridge University Press, 2007.

———. "'The Fashioned Image of Poetry or the Regular Instruction of Philosophy?': Truth, Utility, and the Natural Sciences in Early Modern England." In *Science, Literature, and Rhetoric in Early Modern England,* edited by D. Burchill and J. Cummins, pp. 15–36. Aldershot: Ashgate, 2008.

———. "Laws of Nature in Seventeenth Century England: From Cambridge Platonism to Newtonianism." In *God, Man and the Order of Nature,* ed. Eric Watkins, pp. 127–48. New York: Oxford University Press, 2014.

———. "Natural History." In Harrison, Numbers and Shank, *Wrestling with Nature,* pp. 117–48.

———. "The Natural Philosopher and the Virtues." In Condren, Gaukroger and Hunter, *The Philosopher in Early Modern Europe,* pp. 202–28.

———. "Physico-Theology and the Mixed Sciences: The Role of Theology in Early Modern Natural Philosophy." In *The Science of Nature in the Seventeenth Century,* edited by Peter Anstey and John Schuster, pp. 165–83. Dordrecht: Springer, 2005.

———. "Reading the Passions: The Fall, the Passions, and Dominion over Nature." In *The Soft Underbelly of Reason: The Passions in the Seventeenth Century,* edited by S. Gaukroger, pp. 49–78. London: Routledge, 1998.

———. "Reinterpreting Nature in Early Modern Europe: Natural Philosophy, Biblical Exegesis, and the Contemplative Life." In *The Word and The World: Biblical Exegesis and the Emergence of Modern Science,* edited by K. Killeen and P. Forshaw, pp. 25–44. London: Palgrave Macmillan, 2007.

———. *"Religion" and the Religions in the English Enlightenment.* Cambridge: Cambridge University Press, 1990.

———. "Sentiments of Devotion and Experimental Philosophy in Seventeenth-Century England." *Journal of Medieval and Early Modern Studies* 44 (2014): 113–33.

———. "Subduing the Earth: Genesis 1, Early Modern Science, and the Exploitation of Nature." *Journal of Religion* 79 (1999): 86–109.

———. "Voluntarism and Early Modern Science." *History of Science* 40 (2002): 63–89.

———. "Voluntarism and the Origins of Modern Science: A Reply to John Henry." *History of Science* 47 (2009): 223–31.

———. "Was Newton a Voluntarist?" In *Newton and Newtonianism: New Studies,* edited by James E. Force and Sarah Hutton, pp. 39–64. Dordrecht: Kluwer, 2004.

Harrison, Peter, and David C. Lindberg. "Early Christianity." In Brooke and Numbers, *Science and Religion around the World*, pp. 67–91.

Harrison, Peter, Ronald Numbers, and Michael Shank, eds. *Wrestling with Nature: From Omens to Science*. Chicago: University of Chicago Press, 2011.

Harwood, John. "Rhetoric and Graphics in *Micrographia*." In *Robert Hooke: New Studies*, edited by Michael Hunter and Simon Schaffer, pp. 119–48. Woodbridge: Boydell, 1989.

Hattaway, Michael. "Bacon and 'Knowledge Broken': Limits for Scientific Method." *Journal of the History of Ideas* 39 (1978): 183–97.

Hazlett, Ian. "Calvin's Latin Preface of His Proposed French Edition of Chrysostom's Homilies: Translation and Commentary." In *Humanism and Reform: The Church in Europe, England and Scotland, 1400–1643*, edited by James Kirk, pp. 129–50. Oxford: Oxford University Press, 1991.

Heckel, Martin. "Das Säkularisierungsproblem in der Entwicklung des deutschen Staatskirchenrechts." In *Christentum und modernes Recht: Beiträger zum Problem der Säkularisation*, edited by G. Dilcher and I. Staff, pp. 35–95. Frankfurt am Main: Suhrkamp, 1984.

Heilbron, John. "Natural Philosophy." In Harrison, Numbers, and Shank, eds., *Wrestling with Nature*, pp. 173–99.

Hempton, David. *Methodism: Empire of the Spirit*. New Haven: Yale University Press, 2005.

Henriques, Ursula. *Religious Toleration in England, 1787–1833*. Toronto: University of Toronto Press, 1961.

Henry, John. "Metaphysics and the Origins of Modern Science: Descartes and the Importance of Laws of Nature." *Early Science and Medicine* 9 (2004): 73–114.

———. "Voluntarist Theology at the Origins of Modern Science: A Response to Peter Harrison." *History of Science* 47 (2009): 79–112.

Heyck, T. W. *The Transformation of Intellectual Life in Victorian England*. London: Croom Helm, 1982.

Hibbs, Thomas S. *Dialectic and Narrative in Aquinas: An Interpretation of the Summa Contra Gentiles*. Notre Dame: University of Notre Dame Press, 1995.

Hinde, Robert. *Why Gods Persist: A Scientific Approach to Religion*. London: Routledge, 1999.

Hitchens, Christopher. *God Is Not Great: The Case against Religion*. London: Atlantic Books, 2007.

Holloway, Paul A. "Bona Cogitare: An Epicurean Consolation in Phil 4:8–9." *Harvard Theological Review* 91 (1998): 89–96.

Holmes, Michael W., ed. *The Apostolic Fathers*. Translated by J. B. Lightfoot and J. R. Harmer. 2nd ed. Grand Rapids: Baker, 1989.

Hope, Marjorie, and David Stuart Rodes. Introduction to *The Virtuoso* by Thomas Shadwell, pp. xi–xxvi. Lincoln: University of Nebraska Press, 1966.

Hunter, Michael. *Science and Society in Restoration England*. Cambridge: Cambridge University Press, 1981.

Huntington, Samuel P. *The Clash of Civilizations and the Remaking of World Order.* New York: Simon and Schuster, 1998.

Hutter, H. "Philosophy as Self-Transformation." *Historical Reflections* 16 (1989): 171–98.

Iliffe, Rob. "Abstract Considerations: Disciplines and the Incoherence of Newton's Natural Philosophy." *Studies in History and Philosophy of Science* 35A (2004): 427–54.

Imbach, R. "La Philosophie comme exercice spirituel." *Critique* 41 (1985): 275–83.

Irwin, Terence. "Luther's Attack on Self-Love: The Failure of Pagan Virtue." *Journal of Medieval and Early Modern Studies* 42 (2012): 151–55.

———. *Plato's Ethics.* New York: Oxford University Press, 1995.

Jaeger, Werner. *The Theology of the Early Greek Philosophers.* Oxford: Clarendon, 1948.

Jardine, Nick. "Whigs and Stories: Herbert Butterfield and the Historiography of Science." *History of Science* 41 (2003): 125–40.

Jenkins, John I. *Knowledge and Faith in Aquinas.* Cambridge: Cambridge University Press, 1997.

Jensen, J. Vernon. *Thomas Henry Huxley, Communicating for Science.* London: Associated University Press, 1991.

———. "The X-Club: Fraternity of Victorian Scientists." *British Journal for the History of Science* 5 (1970): 63–72.

Johnson, A. P. *Ethnicity and Argument in Eusebius' "Praeparatio Evangelica."* Oxford: Oxford University Press, 2006.

Jones, Matthew. *The Good Life in the Scientific Revolution: Descartes, Pascal, Leibniz, and the Cultivation of Virtue.* Chicago: University of Chicago Press, 2006.

Jones, Preston. Review of *God Is Not Great* by Christopher Hitchens. *Books and Culture,* April 2007.

Jouanna, Jacques. *Hippocrates.* Translated by M. B. DeBevoise. Baltimore: Johns Hopkins, 1999.

Judge, Edwin. "Was Christianity a Religion?" In *The First Christians in the Roman World,* edited by James R. Harrison, pp. 404–9. Tübingen: Mohr Siebeck, 2008.

Jungkuntz, R. "Fathers, Heretics and Epicureans." *Journal of Ecclesiastical History* 17 (1966): 3–10.

Kaster, Robert A. *Guardians of Language: The Grammarian and Society in Late Antiquity.* Berkeley: University of California Press, 1997.

Kelly, J. N. D. *Early Christian Creeds.* 3rd ed. London: Continuum, 2006.

Kerr, Fergus. *After Aquinas: Visions of Thomism.* Oxford: Blackwell, 2002.

Kiernan, V. "Evangelicalism and the French Revolution." *Past and Present* 1 (1952): 44–56.

King, Richard. "The Association of Religion with Violence: Reflections on a Modern Trope." In *Religion and Violence in South Asia,* edited by John Hinnells and Richard King, pp. 214–42. Abingdon: Routledge, 2007.

Kirby, Richard, Sidney Wirthington, Arthur Burr Darling, and Frederick Gridley Kilgour. *Engineering in History.* New York: Dover Publications, 1990.

Kittel, G., ed. *Theological Dictionary of the New Testament.* 10 vols. Grand Rapids: Eerdmans, 1964–76.

Klee, Robert. *Introduction to the Philosophy of Science: Cutting Nature at Its Seams.* Oxford: Oxford University Press, 1996.

Klueting, Harm. *Das konfessionelle Zeitalter, 1525–1648.* Stuttgart: E. Ulmer, 1989.

Kolbet, Paul R. *Augustine and the Cure of Souls: Revising a Classical Ideal.* Notre Dame: University of Notre Dame Press, 2010.

Korzybski, Alfred. *Science and Sanity: An Introduction to Non-Aristotelian Systems and General Semantics.* Lancaster, PA: International Non-Aristotelian Library, 1941.

Kretzmann, Norman. *The Metaphysics of Creation: Aquinas's Natural Theology in Summa Contra Gentiles.* Oxford: Oxford University Press, 1998.

Krey, Philip D., and Leslie Smith. Introduction to *Nicholas of Lyra: The Sense of Scripture,* edited by Philip D. Krey and Leslie Smith, pp. 1–18. Leiden: Brill, 2000.

Ladner, Gerhart B. *The Idea of Reform: Its Impact on Christian Thought and Action in the Age of the Fathers.* Cambridge, MA: Harvard University Press, 1959.

LaPorte, J. *Natural Kinds and Conceptual Change.* Cambridge: Cambridge University Press, 2004.

Lash, Nicholas. *The Beginning and End of "Religion."* Cambridge: Cambridge University Press, 1996.

Laudan, Larry. *Progress and Its Problems: Towards a Theory of Scientific Growth.* London: Routledge, 1977.

Layton, David. *Science for the People: The Origins of the School Science Curriculum in England.* London: George Allen and Unwin, 1973.

Lehoux, Daryn. "Creation Myths and Epistemic Boundaries." *Spontaneous Generations* 3 (2009): 28–34.

Leiss, William. *The Domination of Nature.* New York: George Braziller, 1972.

Lenzer, Gertrude, ed. *Auguste Comte and Positivism: The Essential Writings.* New York: Harper and Row, 1975.

Lewis, Charles T., and Charles Short. *A Latin Dictionary.* Oxford: Oxford University Press, 1879.

Lieu, Judith. *Christian Identity in the Jewish and Graeco-Roman World.* Oxford: Oxford University Press, 2004.

Lightfoot, J. B. *Commentary on Colossians.* 8th ed. London, 1875.

Lindberg, David C. "The Medieval Church Encounters the Classical Tradition: Saint Augustine, Roger Bacon, and the Handmaiden Metaphor." In Lindberg and Numbers, *When Science and Christianity Meet,* pp. 7–32.

———. "That the Rise of Christianity Was Responsible for the Demise of Ancient Science." In Numbers, *Galileo Goes to Jail,* pp. 8–18.

Little, David. *Religion, Order, and Law: A Study in Pre-Revolutionary England.* New York: Harper and Row, 1969.

Livesey, Steven. "Scientia." In *Medieval Science, Technology, and Medicine: An Encyclopedia,* edited by Thomas Glick, Steven J. Livesey, and Faith Wallis, pp. 455–58. New York: Routledge, 2005.

Lloyd, Claude. "Shadwell and the Virtuosi." *Publications of the Modern Language Association* 44 (1929): 472–94.

Lloyd, G. E. R. *Demystifying Mentalities*. Cambridge: Cambridge University Press, 1990.

———. *Early Greek Science: Thales to Aristotle*. London: Norton, 1970.

———. *In the Grip of Disease: Studies in the Greek Imagination*. Oxford: Oxford University Press, 2003.

———. "Physics." In *The Oxford Companion to Classical Civilization*, edited by Simon Hornblower and Anthony Spawforth, pp. 532–34. Oxford: Oxford University Press, 2004.

Long, A. A., ed. *The Cambridge Companion to Ancient Greek Philosophy*. Cambridge: Cambridge University Press, 1999.

Lottin, O. *Psychologie et morale au XII et XIII siècles*. 7 vols. Louvain-Gembloux: Duculot, 1942–54.

Lovejoy, A. O. *The Great Chain of Being* [1936]. Cambridge, MA: Harvard University Press, 2001.

Lubac, Henri de. *Medieval Exegesis: The Four Senses of Scripture*. Translated by E. M. Macierowski. 3 vols. Grand Rapids: Eerdmans, 1998–2009.

———. *The Splendor of the Church*. San Francisco: Ignatius Press, 1999.

Lukács, Ladislaus, S.J., ed. *Monumenta Paedagogica Societatis Iesu*. 9 vols. Rome: Institutum Historicum Societatis Iesu, 1965–99.

Lüthy, C. "What to Do with Seventeenth-Century Natural Philosophy? A Taxonomic Problem." *Perspectives on Science* 8 (2000): 164–95.

MacDonald, Scott. "Theory of Knowledge." In *The Cambridge Companion to Aquinas*, edited by Norman Kretzmann and Eleanor Stump, pp. 160–95. Cambridge: Cambridge University Press, 1993.

MacIntyre, Alasdair. *After Virtue*. 2nd ed. Notre Dame: University of Notre Dame Press, 1984.

———. *Whose Justice? Which Rationality?* London: Duckworth, 1988.

Mack, Phyllis. *Heart Religion in the British Enlightenment: Gender and Emotion in Early Methodism*. Cambridge: Cambridge University Press, 2008.

Mâle, Emile. *Religious Art in France: The Twelfth Century*. Translated by Marthiel Matthews. Princeton: Princeton University Press, 1978.

Mandair, Arvind. *Religion and the Specter of the West: Sikhism, India, Postcoloniality, and the Politics of Translation*. New York: Columbia University Press, 2009.

Mandelbrote, Scott, and J. van de Meer, eds. *Nature and Scripture in the Abrahamic Religions*. 2 vols. Leiden: Brill, 2008.

Marion, Jean-Luc. *L'Ontologie grise de Descartes*. Paris: Vrin, 1971.

Marsden, George. *Understanding Fundamentalism and Evangelicalism*. Grand Rapids: Eerdmans, 1991.

———. "Understanding Fundamentalist Views of Science." In *Science and Creationism*, edited by A. Montagu, pp. 95–116. Oxford: Oxford University Press, 1984.

Marthaler, Berard L. *The Creed: The Apostolic Faith in Contemporary Theology*. 3rd ed. New London: Twenty-Third Publications, 2007.

Mascall, Eric L. *He Who Is: A Study in Traditional Theism*. London: Darton, Longman and Todd, 1967.

Masuzawa, Tomoko. *The Invention of World Religions*. Chicago: University of Chicago Press, 2005.

McClellan, J., and Harold Dorn. *Science and Technology in World History: An Introduction*. Baltimore: John Hopkins University Press, 1999.

McCutcheon, Russell. "The Category 'Religion' in Recent Publications: A Critical Survey." *Numen* 42 (1995): 285–309.

McDowell, John. "Virtue and Reason." In Crisp and Slote, *Virtue Ethics*, pp. 141–62.

McGrath, Alister. *Iustitia Dei: A History of the Christian Doctrine of Justification*. 3rd ed. Cambridge: Cambridge University Press, 2005.

McMackin Garland, Martha. *Cambridge before Darwin: The Ideal of Liberal Education, 1800–1860*. Cambridge: Cambridge University Press, 1980.

Meer, Jitse van der, and Scott Mandelbrote, eds. *Nature and Scripture in the Abrahamic Religions*. 2 vols. Leiden: Brill, 2008.

Merchant, Carolyn. *The Death of Nature: Women, Ecology, and the Scientific Revolution*. New York: Harper and Row, 1990.

———. *Reinventing Eden: The Fate of Nature in Western Culture*. London: Routledge, 2004.

Merton, Robert K. *The Sociology of Science: Theoretical and Empirical Investigations*. Chicago: University of Chicago Press, 1978.

Methuen, Charlotte. "Interpreting the Books of Nature and Scripture in Medieval and Early Modern Thought: An Introductory Essay." In van der Meer and Mandelbrote, *Nature and Scripture*, vol. 1, pp. 179–218.

Michel, P.-H. "Greek Science." In *Ancient and Medieval Science*, translated by J. Pomerans and edited by R. Taton, pp. 180–242. London, Thames & Hudson, 1963.

Mikkeli, Heikki. *An Aristotelian Response to Renaissance Humanism: Jacopo Zabarella on the Nature of Arts and Sciences*. Helsinki: SHS, 1992.

Milbank, John. *Theology and Social Theory: Beyond Secular Reason*. 2nd ed. Oxford: Wiley-Blackwell, 2006.

Milton, J. R. "Laws of Nature." In *The Cambridge History of Seventeenth Century Philosophy*, edited by D. Garber and M. Ayers, vol. 1, pp. 680–701. 2 vols. Cambridge: Cambridge University Press, 1998.

Mitcham, Carl, and Jim Grote, eds. *Theology and Technology: Essays in Christian Analysis and Exegesis*. New York: University Press of America, 1984.

Molland, A. G. "Medieval Ideas of Scientific Progress." *Journal of the History of Ideas* 39 (1978): 561–77.

Moltmann, Jürgen. *The Theology of Hope*. Minneapolis: Fortress Press, 2012.

Moore, James A. "Cutting Both Ways—Darwin among the Devout: A Response to David Livingstone, Sara Miles, and Mark Noll." *Perspectives on Science and Christian Faith* 45 (1994): 169.

Murray, Gilbert. *Five Stages of Greek Religion*. London: Watts & Co., 1935.

Nagel, Thomas. "Secular Philosophy and the Religious Temperament." In *Secular Philosophy and the Religious Temperament: Essays, 2002–2008*, pp. 3–17. Oxford: Oxford University Press, 2010.

Needham, Rodney. *Belief, Language, and Experience.* Oxford: Blackwell, 1972.

———. *Circumstantial Deliveries.* Berkeley: University of California Press, 1982.

Nehamas, Alexander. *The Art of Living: Socratic Reflections from Plato to Foucault.* Berkeley: University of California Press, 1998.

———. "Socratic Intellectualism." In *Virtues of Authenticity: Essays on Plato and Socrates.* Princeton: Princeton University Press, 1999.

Nichols, Aidan. *The Shape of Catholic Theology: An Introduction to Its Sources.* Edinburgh: T. & T. Clark, 1991.

Nickl, Peter. *Ordnung der Gefühle: Studien zum Begriff des habitus.* Hamburg: Felix Meiner Verlag, 2001.

Niiniluoto, Ilkka. *Is Science Progressive?* Dordrecht: Kluwer, 1984.

Nock, A. D. *Conversion: The Old and the New in Religion from Alexander the Great to Augustine of Hippo.* Oxford: Oxford University Press, 1933.

Noll, Mark. "Evangelicalism and Fundamentalism." In *Science and Religion,* edited by Gary Ferngren, pp. 261–76. Baltimore: Johns Hopkins, 2002.

Nongbri, Brent. *Before Religion: The History of a Modern Concept.* New Haven: Yale University Press, 2012.

———. "Dislodging 'Embedded' Religion: A Brief Note on a Scholarly Trope." *Numen* 55 (2008): 440–60.

Numbers, Ronald L. *The Creationists: From Scientific Creationism to Intelligent Design.* Berkeley: University of California Press, 1993.

———, ed. *Galileo Goes to Jail, and Other Myths about Science and Religion.* Cambridge, MA: Harvard University Press, 2010.

Oakley, Francis. "Christian Theology and the Newtonian Science: The Rise of the Concept of Laws of Nature." *Church History* 30 (1961): 433–57.

O'Connor, W. R. "The *Uti-frui* Distinction in Augustine's Ethics." *Augustinian Studies* 14 (1983): 45–62.

Olsen, Paul. *The Journey to Wisdom: Self-Education in the Patristic and Medieval Literature.* Lincoln: University of Nebraska Press, 1995.

Opelt, I. "Doctrina und doctrina christiana." *Der altsprachliche Unterricht* 9 (1966): 5–22.

Osler, Margaret J. *Divine Will and the Mechanical Philosophy: Gassendi and Descartes on Contingency and Necessity in the Created World.* Cambridge: Cambridge University Press, 1994.

Ott, Walter. *Causation and Laws of Nature in Early Modern Philosophy.* Oxford: Oxford University Press, 2009.

Passmore, John. *Man's Responsibility for Nature: Ecological Problems and Western Traditions.* London: Duckworth, 1974.

Pérez-Ramos, Antonio. "Bacon's Legacy." In *The Cambridge Companion to Bacon,* pp. 680–701. Cambridge: Cambridge University Press, 1996.

Phillips, Denise. *Acolytes of Nature: Defining Natural Science in Germany, 1770–1850.* Chicago: University of Chicago Press, 2012.

Piché, David, ed. *La condemnation parisienne de 1277. Texte latin, traduction, introduction et commentaire.* Paris: Vrin, 1999.

Pinker, Steven. *The Better Angels of Our Nature: Why Violence Has Declined.* New York: Viking, 2011.

Pinsent, Andrew. *The Second Person Perspective in Aquinas's Ethics.* London: Routledge, 2012.

Popper, Karl. "The Myth of the Framework." In *The Myth of the Framework: In Defence of Science and Rationality,* edited by M. A. Notturno, pp. 33–64. London: Routledge, 1996.

Pouillon, J. "Remarks on the Verb 'To Believe.'" In *Between Belief and Transgression: Structural Essays in Religion, History, and Myth,* edited by M. Izard and P. Smith, pp. 1–8. Chicago: University of Chicago Press, 1982.

Preller, Victor. *Divine Science and the Science of God: A Reformulation of Thomas Aquinas.* Princeton: Princeton University Press, 1967.

Press, G. A. "Doctrina in Augustine's *De doctrina christiana.*" *Philosophy and Rhetoric* 17 (1984): 98–120.

Preus, Samuel. "The Reified Heart in the Seventeenth-Century Religion." In *Religion in History: The Word, the Idea, the Reality,* edited by Michel Despland and Gérard Vallée, pp. 45–56. Waterloo: Wilfrid Laurier University Press, 1992.

Proctor, Robert N. *Value-Free Science? Purity and Power in Modern Knowledge.* Cambridge, MA: Harvard University Press, 1991.

Purver, Margery. *The Royal Society: Concept and Creation.* London: Routledge and Kegan Paul, 1967.

Quasten, Johannes. *Patrology.* Vol. 1, *The Beginnings of Patristic Literature.* Westminster, MD: Christian Classics, 1986.

Reichenbach, Hans. *The Rise of Scientific Philosophy.* Berkeley: University of California Press, 1951.

Reuben, Julie A. *The Making of the Modern University.* Chicago: University of Chicago Press, 1996.

Roberts, Jon. "Science and Religion." In Harrison, Numbers, and Shank, *Wrestling with Nature,* pp. 253–79.

Rosenberg, Daniel. "Early Modern Information Overload." *Journal of the History of Ideas* 64 (2003): 1–9.

Ross, Sydney. "'Scientist': The Story of a Word." *Annals of Science* 18 (1962): 65–86.

Rue, Loyal G. *Everybody's Story: Wising Up to the Epic of Evolution.* Albany: SUNY Press, 2000.

Ruel, M. "Christians as Believers." In *Religious Organization and Religious Experience,* edited by J. Davis, pp. 9–32. London: Academic Press, 1982.

Ruler, J. A. van. *The Crisis of Causality: Voetius and Descartes on God, Nature, and Change.* Leiden: Brill, 1995.

Russell, Jeffrey Burton. *Inventing the Flat Earth: Columbus and Modern Historians.* New York: Praeger, 1991.

Russell, Robert J. "Ways of Relating Science and Religion." http://www.counterbalance.org/rjr/atypo-body.html, accessed 15 July 2012.

Russo, Lucio. *The Forgotten Revolution: How Science Was Born in 300BC and Why It Had to Be Reborn*. Dordrecht: Springer, 2004.

Sachot, Maurice. "Comment le christianisme est-il devenu *religio*." *Revue des sciences religieuses* 59 (1985): 95–118.

———. *Quand le christianisme a changé le monde*. Paris: Odile Jacob, 2007.

———. "«Religio/Superstitio», Historique d'une subversion et d'un retournement." *Revue d'histoire des religions* 208 (1991): 355–94.

Samburrsky, S. *The Physical World of the Greeks*. Translated by M. Dagut. London: Routledge, 1956.

Sand, Shlomo. *The Invention of the Land of Israel: From Holy Land to Homeland*. London: Verso, 2012.

Schaffer, Simon. "Scientific Discoveries and the End of Natural Philosophy." *Social Studies of Science* 16 (1986): 387–420.

Schilling, Heinz. *Konfessionskonflikt und Staatsbildung*. Gutersloh: Gütersloher Verlagshaus Mohn, 1981.

Schmidt, Heinrich Richard. *Konfessionalisierung im 16. Jahrhundert*. Munich: Oldenbourg, 1992.

Schmitt, Charles B., and Quentin Skinner, eds. *The Cambridge History of Renaissance Philosophy*. Cambridge: Cambridge University Press, 1988.

Schneewind, Jerome B. "The Misfortunes of Virtue." In Crisp and Slote, *Virtue Ethics*, pp. 178–200.

———. "Philosophical Ideas of Charity: Some Historical Reflections." In *Giving: Western Ideas of Philanthropy*, edited by Jerome B. Schneewind, pp. 54–75. Bloomington: Indiana University Press, 1996.

Scholten, C. *Antike Naturphilosophie und christliche Kosmologie in der Schrift "De opificio mundi" des Johannes Philoponos*. Berlin: De Gruyter, 1996.

Schwartz, Seth. *Imperialism and Jewish Society from 200 B.C.E. to 640 C.E.* Princeton: Princeton University Press, 2001.

Scott, Alan. *Origen and the Life of the Stars*. Oxford: Clarendon, 1991.

Scribner, R. W. *Popular Culture and Popular Movements in Reformation Germany*. London: Hambledon, 1987.

Sedley, David. *Creationism and Its Critics in Antiquity*. Berkeley: University of California Press, 2007.

———. "The Ideal of Godlikeness." In *Plato 2—Ethics, Politics, Religion, and the Soul*, edited by G. Fine, pp. 309–28. Oxford: Oxford University Press, 1999.

———. "Is Aristotle's Teleology Anthropocentric?" *Phronesis* 36 (1991): 179–96.

———. *Plato's Cratylus*. Cambridge: Cambridge University Press, 2003.

Sellars, John. *The Art of Living: The Stoics on the Nature and Function of Philosophy*. Aldershot: Ashgate, 2003.

Seward, Albert C. "Christ's Hospital and the Royal Society." *Notes and Records of the Royal Society of London* 3 (1940–41): 141–45.

Shank, Michael H. "Natural Knowledge in the Latin Middle Ages." In Harrison, Numbers, and Shank, *Wrestling with Nature*, pp. 83–115.

———. "That the Medieval Christian Church Suppressed the Growth of Science." In Numbers, *Galileo Goes to Jail*, pp. 19–27.

Shapin, Steven. *The Scientific Life: A Moral History of a Late Modern Vocation*. Chicago: University of Chicago Press, 2008.

Sillem, Edward. *Ways of Thinking about God: Thomas Aquinas and Some Recent Problems*. London: Darton, Longman and Todd, 1961.

Simmons, Allison J. "Sensible Ends: Latent Teleology in Descartes' Account of Sensation." *Journal of the History of Philosophy* 39 (2001): 49–75.

Simpson, A. D. "Epicureans, Christians, Atheists in the Second Century." *Transactions and Proceedings of the American Philological Association* 72 (1941): 372–81.

Singer, Charles. *A Short History of Science to the Nineteenth Century* [1941]. New York: Dover Publications, 1997.

Skinner, Quentin. "Meaning and Understanding in the History of Ideas." *History and Theory* 8 (1969): 3–53.

Smalley, Beryl. *The Study of the Bible in the Middle Ages*. 2nd ed. New York: Philosophical Library, 1952.

———. *The Study of the Bible in the Middle Ages*. 3rd ed. Oxford: Blackwell, 1983.

Smith, Jonathan Z. *Map Is Not Territory: Studies in the History of Religions*. Leiden: Brill, 1978.

Smith, Wilfred Cantwell. *Faith and Belief: The Difference between Them*. Princeton, NJ: Princeton University Press, 1979.

———. *The Meaning and End of Religion* [1962]. London: SPCK, 1978.

Sorabji, Richard, ed. *Philoponus and the Rejection of Aristotelian Science*. Ithaca: Cornell University Press, 1987.

———. *The Philosophy of the Commentators, 200–600 AD*. Vol. 2, *Physics*. London: Duckworth, 2003.

Sorell, Tom, G. A. J. Rogers, and Jill Kraye, eds. *Scientia in Early Modern Philosophy*. Dordrecht: Springer, 2010.

Soulen, Kendall. *The God of Israel and Christian Theology*. Minneapolis: Fortress Press, 1996.

Spring, David and Eileen, eds. *Ecology and Religion in History*. New York: Harper and Row, 1974.

Steinle, F. "The Amalgamation of a Concept—Laws of Nature in the New Sciences." In *Laws of Nature: Essays on the Philosophical, Scientific and Historical Dimensions*, edited by F. Weinert, pp. 316–68. Berlin: De Gruyter, 1995.

Steinmetz, David C. "What Luther Got Wrong." *Christian Century*, 23 August 2005, 23–25.

Stenmark, Mikael. "Ways of Relating Science and Religion." In *The Cambridge Companion to Science and Religion*, edited by Peter Harrison, pp. 278–94. Cambridge: Cambridge University Press, 2010.

Stewart, Larry. *The Rise of Public Science: Rhetoric, Technology, and Natural Philosophy in Newtonian Britain, 1660–1750*. Cambridge: Cambridge University Press, 1992.

Stock, Brian. *After Augustine: The Meditative Reader and the Text*. Philadelphia: University of Pennsylvania Press, 2001.

———. *Augustine the Reader: Meditation, Self-Knowledge and the Ethics of Interpretation*. Cambridge, MA: Harvard University Press, 1996.

Stone, M. W. F. "The Adoption and Rejection of Aristotelian Moral Philosophy in Reformed 'Casuistry.'" In *Humanism and Early Modern Philosophy*, edited by Jill Kraye and M. W. F. Stone, pp. 59–90. London: Routledge, 2000.

Stroumsa, Guy. *The End of Sacrifice: Religious Transformations in Late Antiquity*. Translated by Susan Emanuel. Chicago: University of Chicago Press, 2009.

———. *A New Science: The Discovery of Religion in the Age of Reason*. Cambridge, MA: Harvard University Press, 2010.

Stump, Eleonore. *Aquinas*. London: Routledge, 2003.

———. "Aquinas on the Foundations of Knowledge." In *Aristotle and His Medieval Interpreters*, edited Richard Bosley and Martin Tweedale, pp. 125–58. *Canadian Journal of Philosophy* Supplementary Volume 17 (Calgary, 1991).

Syfret, R. H. "Some Early Critics of the Royal Society." *Notes and Records of the Royal Society of London* 8 (1950): 20–64.

Tanzella-Nitti, G. "The Two Books Prior to the Scientific Revolution." *Annales Theologici* 18 (2004): 51–83.

Taub, Liba. *Ptolemy's Universe: The Natural Philosophical and Ethical Foundations of Ptolemy's Astronomy*. Chicago and La Salle, IL: Open Court Press, 1993.

Taylor, Charles. *A Secular Age*. Cambridge, MA: Harvard University Press, 2007.

———. "The Politics of Recognition." In *Multiculturalism: Examining the Politics of Recognition*, ed. Amy Gutman, pp. 25–73. Princeton: Princeton University Press, 1994.

Thiessen, Matthew. *Contesting Conversion: Genealogy, Circumcision, and Identity in Ancient Judaism and Christianity*. Oxford: Oxford University Press, 2011.

Thomassen, Beroald. *Metaphysik als Lebensform: Untersuchungen zur Grundlegung der Metaphysik im Metaphysikkommentar Alberts des Grossen*. Munster: Aschendorff, 1985.

Thurs, Daniel P. "Scientific Methods." In Harrison, Numbers, and Shank, *Wrestling with Nature*, pp. 307–35.

Todorov, Tzvetan. *The Fear of Barbarians: Beyond the Clash of Civilizations*. Chicago: University of Chicago Press, 2010.

Topham, Jonathan. "Beyond the 'Common Context': The Production and Reading of the Bridgewater Treatises." *Isis* 89 (1998): 233–62.

———. "Science and Popular Education in the 1830s: The Role of the *Bridgewater Treatises*." *British Journal for the History of Science* 25 (1992): 397–430.

Turner, Frank. "The Victorian Conflict between Science and Religion: A Professional Dimension." *Isis* 49 (1978): 356–76.

Veyne, Paul. *Did the Greeks Believe in Their Myths?* Translated by Paula Wissing. Chicago: University of Chicago Press, 1988.

Vlastos, Gregory. "Theology and Philosophy in Early Greek Thought." In *The Presocrat-

ics, edited by Daniel W. Graham, pp. 3–31. Princeton: Princeton University Press, 1993.

Voelke, André-Jean. *La philosophie comme thérapie de l'âme.* Paris: Cerf, 1993.

Wallace, William A. *Galileo and His Sources: The Heritage of the Collegio Romano in Galileo's Science.* Princeton: Princeton University Press, 1984.

Walsh, J. J. "On Christian Atheism." *Vigiliae Christianae* 45 (1991): 255–77.

Waterfield, Robin. *The First Philosophers: The Presocratics and Sophists.* Oxford: Oxford University Press, 2000.

Webster, Charles. "The Authorship and Significance of Macaria." *Past and Present* 56 (1972): 34–48.

———. *The Great Instauration: Science, Medicine and Reform, 1626–1660.* New York: Holmes & Meier, 1976.

Weijers, Oga. "L'appellation des disciplines dans les classifications des sciences aux XIIᵉ et XIIIᵉ siècles." *Archivum Latinitatis Mediae Aevi* 46/47 (1988): 42–43.

Weinberg, Steven. *The First Three Minutes.* New York: Basic Books, 1977.

Weldon, Stephen. "In Defense of Science: Secular Intellectuals and the Failure of Nerve Thesis." *Religious Humanism* 30 (1996): 30–39.

White, Lynn, Jr. "The Historical Roots of Our Ecological Crisis." *Science* 155, no. 3767 (1967): 1203–7.

Whitney, Elspeth. "Lynn White, Ecotheology, and History." *Environmental Ethics* 15 (1993): 151–69.

Wildberg, C. "Impetus Theory and the Hermeneutics of Science in Simplicius and Philoponus." *Hyperboreus* 5 (1999): 107–24.

Wilson, David Sloan. *Darwin's Cathedral: Evolution, Religion and the Nature of Society.* Chicago: University of Chicago Press, 2002.

Wilson, E. O. *The Creation: An Appeal to Save Life on Earth.* New York: Norton, 2006.

———. *On Human Nature.* Cambridge, MA: Harvard University Press, 1978.

Wilson, Robert. *Astronomy through the Ages.* Princeton: Princeton University Press, 1997.

Wippel, John F. "The Condemnations of 1270 and 1277 at Paris." *Journal of Medieval and Renaissance Studies* 7 (1977): 169–201.

Wittgenstein, Ludwig. *Tractatus Logico-philosophicus.* London: Routledge, 1994.

Wolfson, Harry. "Nicolaus of Autrecourt and Ghazali's Argument against Causality." *Speculum* 44 (1969): 234–38.

Wolpert, Lewis. "The Unnatural Nature of Science." In *Unveiling the Microcosmos: Essays on Science and Technology from the Royal Institution,* edited by Peter Day, pp. 143–56. Oxford: Oxford University Press, 1996.

Yasukata, Toshimasa. *Lessing's Philosophy of Religion and the German Enlightenment.* New York: Oxford University Press, 2002.

Yeo, Richard. *Defining Science: William Whewell, Natural Knowledge and Public Debate in Early Victorian Britain.* Cambridge: Cambridge University Press, 1993.

———. "An Idol of the Market-Place: Baconianism in Nineteenth Century Britain." *History of Science* 23 (1985): 251–98.

———. "The Principle of Plenitude and Natural Theology in Nineteenth-Century Britain." *British Journal for the History of Science* 19 (1986): 263–82.

———. "Reading Strategies for Coping with Information Overload, ca. 1550–1700." *Journal of the History of Ideas* 64 (2003): 11–28.

———. "William Whewell, Natural Theology and the Philosophy of Science." *Annals of Science* 26 (1979): 493–512.

Yolton, John. "Schoolmen, Logic and Philosophy." In *The History of the University of Oxford*. Vol. 5, *The Eighteenth Century,* edited by L. S. Sutherland and L. G. Mitchell, pp. 565–90. Oxford: Clarendon, 1986.

Young, Frances. *Biblical Exegesis and the Formation of Christian Culture.* Cambridge: Cambridge University Press, 1997.

———. "The Rhetorical Schools and Their Influence on Patristic Exegesis." In *The Making of Orthodoxy: Essays in Honour of Henry Chadwick,* edited by Rowan Williams, pp. 182–99. Cambridge: Cambridge University Press, 1989.

Young, Robert M. *Darwin's Metaphor: Nature's Place in Victorian Culture.* Cambridge: Cambridge University Press, 1985.

[INDEX]

Page numbers followed by *f* indicate a figure.

Plato: on a divinely ordered cosmos, 25, 28, 45, 76–77, 212n28; on the goal of philosophy, 28, 33; on good concepts, 6; religious temperament of, 55–56; taxonomy of knowledge of, 31, 41–42; on *theologia*, 17, 208n49; on use of charms, 46, 219n106

Plattes, Gabriel, 237n59

Plotinus, 229n87

Pluche, Noël-Antoine, 149

Plutarch, 212n39, 221n134

Pocock, John, 184

Politics (Aristotle), 63–64

Pope, Alexander, 123–24

Popper, Karl, 24, 39, 52

Porphyry of Tyre, 32, 39

positivism, 118, 143, 174–75, 190–91

Posterior Analytics (Aristotle), 13–14, 147

Preliminary Discourse to the Study of Natural Philosophy (Herschel), 152

pre-Socratic thought, 24–25, 211n12

Preus, Samuel, 235n46

Principia Mathematica (Newton), 109–10, 186

principle of plenitude, 224n22

Principles and Duties of Natural Religion (Wilkins), 150

The Principles of Moral and Political Philosophy (Paley), 149–50

Principles of Science: A Treatise on Logic and the Scientific Method (Jevons), 168

Proast, Jonas, 238n71

professionalization of science, 148, 159–64

progress (as an idea in the early modern periood), 117–44, 174–75, 251n87; accumulation of experimental knowledge and, 117–24, 241nn14–15; charity and, 119, 131–36, 141, 143, 243nn49–50; dominion over nature in, 119, 136–41, 243nn60–61; the Fall and, 188, 252n9; printed sources in, 123–24; redefined virtues and values of, 119, 135, 143–44, 245n75; on religion's inhibiting role in, 141–43; religious and moral usefulness of, 127–31, 135–36; relocation from personal to historical realm of, 120–21; satire of, 125–26; utility for the common good of, 124–31, 181

The Protestant Ethic and the Spirit of Capitalism (Weber), 190–91

Protestant Reformation. *See* Reformation era

Ptolemy, Claudius, 33, 181

Pufendorf, Samuel, 135, 218n96

Puritans, 98–99, 142, 235n46

Rabanus Maurus, 223n9

Radulfus Ardens, 9

Rational Account of the Grounds of the Protestant Religion (Stillingfleet), 107–8

Ray, John, 78

Raymond of Sabunde, 73

Reasonableness and Certainty of the Christian Religion (Jenkin), 150

Reasons of the Christian Religion (Baxter), 106

Recorde, Robert, 81

Rees, Martin, 179

Reformation era, 11, 85; censure of allegory and exegesis in, 75–76, 228n71; critique of Aristotelian-Thomist teleology in, 84–92, 185–86, 188, 231n9, 233n26; on discipline, 96; explicit faith and the definite article in, 92–94, 98, 131, 147, 234nn31–32; on the Fall and original sin, 86–88, 188, 252n9; on grace and divine will, 79, 86–87; instruction of the laity in, 94–96; motivation for experimentation in, 80–81, 88–91; natural theology of, 113–15, 127–29, 148–49, 175, 188–93; physico-theology of, 110–13, 157, 190; plural understanding of "religions" in, 97–103, 147, 176, 235n43; political constructs of religion of, 97–98, 99, 100–101f, 116, 254n27; rational justification of faith in, 105–8, 131; on salvation by good works, 85–87, 231n9; scientific progress and, 142–43; on the true goals of human life, 89–90, 233nn21–23; on true religion, 92–94, 102–8, 236n47, 236nn51–52. *See also* progress of the early modern period

Reichenbach, Hans, 253n21

religio, x; alternative terminology for, 17–18, 208nn47–50; exterior objectification of, 11, 15–16, 74, 84–92, 96–97, 120–21, 141–42, 187; as interior disposition, 7–11,